I0834908

Carefully prepared instructions, in detail, for setting up and running, are sent with each machine, so that, with ordinary tact and ability, they may be set up, and operations commenced, without difficulty or hinderance.

A set of machinery occupies a space of about four by fifteen feet, requires about ten-horse power to run it, with twelve men, three boys, and two horses, to cut the crude material from the bog, haul it to the works, tend the engine and machinery, and place the wet blocks of peat, as they come from the machine, on to the spreading-ground, to dry.

The machine receives the crude peat just as it is taken from the bog, *condenses* it, and, in a very few minutes, delivers it in the form of bricks, which may then be exposed in the open air or under shelter, on the ground, or on a flooring of boards, or on frames prepared for the purpose, to dry and cure. The time ordinarily required, in the summer season, for drying sufficiently to be housed, is about six or eight days, though it varies from four to ten, according to the weather.

The process is exceedingly simple, and the machinery equally so. The original organization of the peat is entirely destroyed; the air (of which a large amount is contained in its cells) is ejected; advantage is taken of some of the natural properties of the material, and the mass is condensed in its moist state; in which condition it is formed into blocks of convenient size and shape, and delivered from the machine to be exposed for drying, as before mentioned, in much the same manner as bricks are exposed in a brick-yard.

The fuel prepared by this process is called *condensed* peat, in contradistinction to *compressed* peat; the material being absolutely *condensed*, without employing any very considerable *pressure* in the process.

The buildings for sheltering the machinery and housing the fuel should be located on dry ground, near to and by the side of the bog, and may be of the most inexpensive character.

For one set of machinery, with engine and boiler, a building twenty by twenty feet is large enough; and, for each additional set of machinery, ten feet in length should be added. The most desirable location for this building is on a side-hill; so that, while the machinery is placed on the lower floor, the crude material may be easily hauled to the level of the second floor, and there dumped by the cart-load into the hopper prepared to receive it. The height between floors in this building should be ten feet.

The other buildings required are simply for shelter of the fuel after it is sufficiently dry to be housed; and their extent must, of course, be regulated by the amount manufactured.

The laborers employed, with the exception of an engineer, may be of the least expensive class.

Peat varies in its character very materially in different localities; but, as a general thing, we estimate that it is reduced in the process of manufacture about two thirds to three quarters, both in weight and bulk, according to the character or composition of the crude material, and the drainage of the meadow or bog from which it is cut. Peat from a well-drained meadow, retaining, of course, less water in the mass, is much more conveniently and economically manufactured than that from a meadow which is constantly overflowed; and the shrinkage, it will readily be understood, is less.

From the best peats we can, of course, produce a superior article of hard, dry fuel; but an essential feature of our process is, that we are able to produce, with equal ease, an excellent fuel from inferior and comparatively worthless crude material.

The cost of one set of machinery, of 100 tons capacity, together with the needful buildings, engine, boiler, shafting, &c., may be anywhere from $4000 to 6000, and the cost of an establishment with a fifty ton machine, is ordinarily about a thousand dollars less.

One set of machinery turns out the material for twenty-five tons, or more, of dry fuel per day, of value, at the present time, of $10 per ton, which is equal to $250 per day. The cost of labor to produce this is less than $2 per ton, to which may be added as much more for cost of transportation to market, say, whole cost delivered in market $4 per ton, and this shows a net gain of $150 on the product of each day.

If, then, out of the season, which is from April to October, seven months, we run only one hundred days, the gain will amount to $15,000; or if one hundred and fifty days, it will amount to $22,500. If, however, we reduce the price even as low as $6 per ton, the gain in one hundred days will be $5000, or, in one hundred and fifty days, it will be $7500; in either case it shows a business of no inconsiderable gain, especially when we take into consideration the small amount of capital required to conduct it.

The cost, as above stated, is for the product of one set of machinery: but, where several machines are to be operated in one establishment, and the business is to be conducted on an extensive scale, the cost *pro rata* will be very much reduced, as one man can easily superintend several machines, the laborers generally can be employed to better advantage, and numerous mechanical appliances to save manual labor and expedite the operations, which it would not be advisable to construct where a single machine only was to be run, may be economically introduced on more extensive works; and in this manner it is quite probable that the expense may be reduced to one dollar and twenty-five cents per ton, and some have estimated as low as one dollar per ton. The profit of such an establishment will therefore be largely in excess of smaller works.

We build machines only to order, and for this reason it is desirable that parties wishing to purchase should give their orders in ample season, so as to avoid the possible delay which might otherwise occur.

☞ Our Works are at East Lexington, about ten miles from Boston, and are freely open to inspection.

Cars leave the depot of the Fitchburg Railroad, in Boston, at 8.35, A. M., 12, M., and 2.45, P. M. Get out at Pierce's Bridge Station. Returning, leave Lexington at 1.22, 3.40, and 6.10, P. M.

A full-sized machine can be seen in motion, in a building near our office in Boston.

☞ RIGHTS FOR TOWNS, COUNTIES, AND STATES FOR SALE.

LEAVITT & HUNNEWELL,

AGENTS FOR THE BOSTON PEAT CO.,

No. 49 Congress Street, Boston.

We are represented by

Messrs. PARDEE & BENEDICT, 31 Pine Street, New York;

ELISHA CONGDON, Esq., Chelsea, Michigan;

Messrs. DELAPLAINE, BURDICK & GRAY, Madison, Wis.; to either of whom application may be made for information, or for the purchase of rights or machines.

LEAVITT'S PEAT JOURNAL:

A NEWSPAPER,

DEVOTED ESPECIALLY TO THE DISSEMINATION OF INFORMATION RELATING TO

PEAT,

ITS PREPARATION AND USE

AS AN ARTICLE OF FUEL,

AND GENERALLY TO ALL THAT PERTAINS TO

The Economical Production and Use of Fuel

OF ALL KINDS.

ISSUED MONTHLY,

BY

LEAVITT & HUNNEWELL,

49 Congress St., Boston.

PRICE FIFTY CTS. PER ANNUM.

For Sale by all Newsmen.

LEAVITT'S PEAT CONDENSING AND MOULDING MILL.

FACTS ABOUT

PEAT

AS AN ARTICLE OF FUEL.

WITH

REMARKS UPON ITS ORIGIN AND COMPOSITION, THE LOCALITIES IN WHICH IT IS FOUND, THE METHODS OF PREPARATION AND MANUFACTURE, AND THE VARIOUS USES TO WHICH IT IS APPLICABLE; TOGETHER WITH MANY OTHER MATTERS OF PRACTICAL AND SCIENTIFIC INTEREST.

TO WHICH IS ADDED, A CHAPTER ON THE

UTILIZATION OF COAL DUST WITH PEAT,

FOR THE PRODUCTION OF AN EXCELLENT FUEL, AT MODERATE COST,

SPECIALLY ADAPTED FOR STEAM SERVICE.

T. H. LEAVITT.

THIRD EDITION,—REVISED AND ENLARGED.

BOSTON:
LEE AND SHEPARD.
1867.

Stereotyped at the Boston Stereotype Foundry,
4 Spring Lane.

CONTENTS.

PAGE

INTRODUCTION TO THIRD EDITION 7

PRELIMINARY REMARKS 9

Importance attached to fire; its universal use; full value and influence upon changes in nature; heat, light, and fuel essential to man; artificial heat required in useful arts; its use as motive-power comparatively modern; influence of fuel upon the political power of a country; material sources of heat; substances employed as fuel; fuel a source of wealth; what substances are termed fuel; when inflammable fuels are required, and how peat answers such requirements; wood, peat, and coal allied; changes induced by physical effects; analogy existing between peat and coal; vegetable origin; peat as fuel comparatively unknown; value of peat not realized; aim of former manufacturers; numerous machines, but no satisfactory method of manufacture; failure through lack of comprehension of nature of peat; popular error in manufacture of peat; simplicity and cheapness of recent inventions; superiority of peat for domestic and manufacturing purposes; peat for generating steam; can compete in cost with coal; peat for railroads; every manufactory can have its own peat-meadow, and the result of such ownership; Professor Emmons's remarks and opinions.

ORIGIN AND COMPOSITION OF PEAT 20

What is it? color and consistency; natural history; theories of formation; peat in England, Ireland, Scotland, Germany, Holland, and France; area, depth, and quality; stratification with gravel, clay, shells, &c.; opinions of Professor Dana, Macculloch, Williams, Sir James Hall, Dr. Rennie, and others; "recent peat" and "older peat;" peat-meadows may be probed, and how; density; percentage of moisture; highly inflammable; animal and vegetable remains found in peat-bogs; percentage of ash; odor from burning peat; peat peculiar to cold climates; peat in England, Scotland, and France; Professor Lyell and Dr. Darwin quoted; early mention of peat from various authorities; sweet and wholesome fire for invalids.

METHODS OF PREPARATION FOR FUEL 32

Primitive methods of cutting and preparing peat-fuel; peat-knife or *slane* described; recent operations at Barnstable; peat sold in Boston; "hand-peat" in Holland and Ireland; machines for cutting peat; dredging peat; *best* peat-cutter, and what it will accomplish; B. H. Paul on "The Utilization of Peat;" method in Scotland; methods of manufacture in Europe, devised by Williams, Codbold, Stones, Buckland, Hodgson, Gwynne & Co., Mannhardt, Exter, Versmann, Challeton, Siemens, Weber, Gysser, Schlickensen, and others, comprising the *wet* process and *dry* process; advantages and disadvantages of the two processes; methods of manufacture in America; variety of presses and machines; description of works at Lexington, Mass., Pekin, N. Y., Bulstrode, C. W., Belleville, N. J., comprising the Ashcroft & Betteley, Roberts, Hodges, Elsberg, and Leavitt machines; estimates for putting up peat-works; cost of labor to manufacture peat-fuel; data concerning peat-beds, and results of observation from practical operations at the works of the Boston Peat Co. at Lexington; method of working; yield of dry fuel; specific gravity; weight per cord; quality improves by age; should be housed; drying peat; peculiarities; what is required; steady progress is being made.

PEAT-CHARCOAL 69

Simple mode of carbonizing; advantage gained by coking; density superior to best wood-charcoal; calorific power intense; specimens in Great Exhibition at Paris in 1851; peat and peat-charcoal used in Paris; general principles of carbonizing the same in all countries; Professor Johnson on its value and uses; comparative value of charcoal from peat manufactured by the *dry* and *wet* processes, with tests; high heating power; disinfecting and fertilizing qualities.

PEAT IN EUROPE 73

Ireland; Mr. Griffith's description of bogs of Ireland; bog of Dourah; how bog-timber is found; "The Bog of Allen;" extent of peat soil in Ireland; value of amount cut, and importance of peat in Ireland; Mr. Aher quoted; forests overthrown and covered with peat; timber, coins, arms, and utensils found in peat; forests destroyed; bog-iron ore.

PEAT IN FRANCE 81

Peat-fuel for Paris market; where cut; quantity; value; peat-charcoal and ashes; "best fuel the earth produces;" French treatise on peat; uses of peat and charcoal; value of peat-ashes; estimates of the value of peat-fuel by French engineers; consumption of fuel of all kinds in France; its equivalent in peat; cost of working pig iron; economy of peat for *all* purposes; estimate of peat-fuel required for the entire locomotive service of France; bulk of peat-fuel; testimony of Supt. of R. R. Engineers; density of peat-fuel.

PEAT IN ITALY 87

Successful introduction of peat in the puddling of iron and steel; accomplished by the Siemens Gas Furnace; iron mines of Elba; amount of pig iron produced.

PEAT ON THE FALKLAND ISLANDS 88

Destitute of coal and wood; peat is abundant.

PEAT IN NEWFOUNDLAND 88

Large quantities of good peat; the manufacture of it commenced.

PEAT IN NOVA SCOTIA 88

Peat-bogs numerous; well decomposed and bituminous.

PEAT IN ASIA 89

Major Risley's account of "combustible mud;" tried on locomotive at Cawnpore; compared with wood; cost; superstitions of the natives.

PEAT IN CANADA 90

Location, character, and extent of deposits of peat; Mr. Hodges's operations; experiments on Grand Trunk R. R.; contract to furnish peat to G. T. R. R. Co.; importance of the fuel question in Canada; estimate of annual amount to be saved by the use of peat; government should aid the peat-fuel enterprise; probabilities of sufficient supply; present suffering and privation for lack of cheap fuel; supply abundant; requires only enterprise and capital.

PEAT IN MAINE 101

Statements by Dr. Jackson; analysis; similarity to coal; theory of its formation; superiority over wood and coal; abundant deposits; localities; may be used for burning lime and for domestic purposes; Campo Bello.

PEAT IN NEW HAMPSHIRE 103

Deposits numerous and of excellent quality; localities; prospects.

PEAT IN VERMONT 104

Peat-beds in all parts of the State; great depth; interesting peculiarities; localities; marl-beds; beaver works; remarkable discoveries;

bones of elephants found in peat; peat suitable for gas; peat at great elevation.

PEAT IN MASSACHUSETTS 108

Dr. Hitchcock's statements and opinions; eighty thousand acres of peat in the State; peat fever; localities; varieties of peat; peculiarities of peat-beds; submerged forests; letter from Elias Phinney; peat-fuel as compared with wood; formation of peat; when and how to cut and cure; its value; Dr. Rennie on peat; peat-water astringent and antiseptic; salubrity and healthfulness of peat-bogs; similarity of peat and coal; similarity of origin.

PEAT IN RHODE ISLAND 117

Peat in nearly every town in the State; some of excellent quality, and is to be worked for fuel; manufacturing interests promoted by it.

PEAT IN CONNECTICUT 118

Rich in deposits of peat; Professor Johnson's report; localities; preparations for producing the fuel; its importance to manufacturing establishments and railroads.

PEAT IN NEW YORK 120

Reports by Professor Mather, Mr. Vanuxem, and Dr. Emmons; high economical value of peat; extracts from reports; importance urged; great body of heat; substitute for best Liverpool coal; how to test it; where to find it; ranks next to coal for sustaining high temperature; abundance and cheapness commend it to public attention; extended list of localities; bones of the mastodon found in peat in several localities; ostrich egg in peat; numerous peat-beds on Long Island; peat is an important item in the natural resources of wealth in the State.

PEAT IN NEW JERSEY 138

Deposits numerous; interest manifested; localities; marl beds; cedar swamps, and what is found in them; great age of some of the trees; peat very pure; large amount of timber raised from the swamps; used for rails and shingles; immense bogs.

PEAT IN PENNSYLVANIA 141

Peat-bogs are unripe beds of coal; interesting and valuable extracts from Geological Report of the State; description of a very peculiar peat-bog in Denmark; Dismal Swamp; the beautiful economy of nature; its laws are simple; progress slow; results wonderful; remarkable bed of lignite in Germany; another peculiar deposit in Denmark.

PEAT IN VIRGINIA 146

The Great Dismal Swamp; peat companies to operate in it; "Bare Garden;" peculiarities of growth and formation; analyses of peats by Professors Johnson and Silliman; formation of these deposits; quality and depth of peat; Dismal Swamp Canal; peculiarities of the wood, water, and atmosphere of the swamp; choice timber; rich and pure peat; facilities for manufacture and shipment; area and depth; capital and enterprise attracted; sweet water; antiseptic properties.

PEAT IN OHIO 156

Information wanted.

PEAT IN MICHIGAN 156

Letter from Professor Winchell; deposits numerous, deep, and of good quality; importance of the subject to housekeeper, capitalist, and manufacturer; analyses; supply inexhaustible; operations commenced.

PEAT IN INDIANA 159

Thousands of acres of good quality; operations commenced; one bog containing one hundred and eighty square miles of peat, forty feet deep; ample supply along the lines of railroads.

PEAT IN ILLINOIS 161

Valuable deposits; supply of coal; consumption of coal; high cost of coal; importance of the fuel question; peat and its uses for domestic purposes, manufacturing and locomotive service; advantages to result from the introduction of peat-fuel; "cheap fuel" will make Chicago a great *manufacturing* city; explorations and anticipations.

PEAT IN WISCONSIN. 164

Peat is abundant; wood and coal scarce; characteristics of peat-beds in this region; Dr. Hayes's report on Wisconsin peat; calorific power and illuminating properties; the "Northern Farmer" on peat; important source of wealth; more valuable than coal; cost of manufacture; saves the timber; worthless marsh lands are now valuable as coal-beds; money and enterprise are invited.

PEAT IN IOWA 168

Importance of the subject; State Geologist is investigating; lack of fuel is a great hinderance to the settlement of the State; peat offers to remedy the difficulty; letter from Dr. White, State Geologist; formation of peat; its value; successful method and machinery for preparing it; lack of timber and coal; explorations for peat to be continued.

PEAT IN MINNESOTA 170

What the "St. Paul Press" says; exhaustless deposits of peat in every Northern State; Minnesota "beats all creation" on peat; peat and petroleum; peat machines; green bass-wood at eight dollars per cord, and a hint to "enterprising citizens."

KANSAS, NEBRASKA, COLORADO, UTAH, AND NEVADA 173

Rumors and reports of extensive peat deposits; government orders to explore, investigate, and report.

PEAT IN CALIFORNIA 173

Tula marshes identical with peat; inquiries instituted.

PEAT IN THE MANUFACTURE OF IRON 173

Peat can be used economically and profitably; early experience with coal; peat has been successfully employed in England, France, Italy, Bohemia, Bavaria, Westphalia, Wurtemberg, and elsewhere; peat-coke is of greater value than best wood-charcoal. Peat preferable for welding, fine iron work, &c.; can be used in puddling and reverberatory furnaces and forges, and in blast furnaces; comparative heating power; advantage gained by condensing peat; particulars concerning the use of peat in iron works at Konigsbronn, Ichoux, Ransko, Schlackenwerth, Weiherhammer, Magdeberg, Nerecia, and in Italy; comparative cost and results; importance of the subject to France, Germany, and Sweden; extracts from published statements; iron made with peat charcoal will not splinter; peat used in smelting lead; gas furnaces for the treatment of metals and glass; description of the Siemens Regenerative Gas Furnace, and the practicability of employing peat and other cheap fuels in it; advantages to accrue to certain sections from the adoption of this process; establishments now using it; W. E. Newton before Soc. of Arts; experiments and results; D. K. Clark, C. E., before Brit. Asso.; further experiments, results, and opinions; iron metallurgists are agreed; tests in Sweden and Germany, at Bolton and Horwich; reports of Mr. Sanderson of Sheffield, Mr. Fothergill of Oldham, Messrs. Brown and Lennox, Professor Emmons, Mr. McDougall, Mr. Campbell, Professor Johnson, and Lond. Mech. Mag., embodying tests, experiments, opinions, and results; economical use of peat; peat in Austria in puddling, reheating, and blast furnaces.

PEAT AS APPLIED FOR GENERATING STEAM . . . 198

Use of peat on steamers; Mr. Williams's patents; peat for fireworks, and for softening steel; Mr. Brunton, Mr. Paul, and Professor Emmons; statements and illustrations; Mr. Nursey before Soc. of Engineers, on English coal fields; deposits of peat in Great Britain and Ireland; cost of peat-fuel; qualities superior to coal; experiments on steamboat and locomotive, in smelting iron, in puddling furnace, and for producing gas; Lond. Mech. Mag.; statements and opinions; peat is the only fuel for locomotives in South Bavaria; used on an English railway; test at the Horwich works; trial on Paris and Lyons R. R.; no smoke, much gas, constant flame; advantages of peat for steam-fuel enumerated; saving of grate-bars and furnaces; trials by Nar. Brick Co. on N. Y. Central, Hartford, and Spr., Hud. River, Eastern, and Vt. Central, and Grand Trunk railroads, with details of cost and results; trials at East Boston and Lowell; the fuel is proved to be "*good;*" "the people have to learn *how to use it.*"

PEAT FOR DOMESTIC PURPOSES 218

Importance of observing *how* to use it; methods and appliances for using it economically and advantageously; statements and opinions; peat used in Boston and vicinity; peat for kindling; peat sold in boxes like wood, and in bulk like coal.

INTENSITY OF HEAT GENERATED BY PEAT 221

Intense heating properties are of special importance; comparative composition of fuels; comparative importance of *quantity* and *intensity* of heat; reasons for converting coal into coke, and wood into charcoal.

HOW TO USE PEAT 223

Peats differ in quality and characteristics in same manner as woods and coals; stoves, furnaces, grates, drafts, &c., to be adapted; appliances required are simple; for locomotives and stationary engines; use of fuel dry and partially dried; progress of economy in the use of fuels generally.

GAS FROM PEAT 230

Tested in Europe; yield equal to coal; brilliancy and power superior; early experiments; gas furnaces for iron in France, Germany, Prussia, and Sweden; used at Dartmoor prison; high illuminating power of peat gas and peat oil compared with coal gas; statements by Mr. Paul, Mr. Keats, Mr. Brunton, Mr. Versmann, and Professor Emmons; advantages of peat for gas; importance for gas second only to coal for fuel; tests by Dr. Hayes; experiments at Portland, Utica, Lansingburg, &c.; retorts for producing it; quantity, density, and illuminating power as compared with gas from coal.

PEAT IN GUNPOWDER AND FIREWORKS 239

Superior to charcoal from dog-wood and alder; combustion more instantaneous and perfect; used in Europe for brilliant colored fires.

CHEMICAL PRODUCTS FROM THE DISTILLATION OF PEAT 240

Statements by Lord Ashley in 1849; tests by Dr. Hodges; estimate by Coffey & Sons, for producing on a large scale; expenditures and returns; Professor Brande's opinions; products of destructive distillation; bleaching properties of peat-charcoal; Irish Peat Company's operations; B. H. Paul on the manufacture of hydro-carbon oils, paraffine, &c., from peat; analine colors and paraffine candles from peat.

ANALYSES AND PROPERTIES OF PEAT 247

Ultimate elements; chemical products; reports of analyses; contents of carbon, hydrogen, oxygen, nitrogen; percentage of ashes; specific gravity and composition of peat ashes; methods of analyses and distillation; water, tar, charcoal, gas; ammonia, acetic acid, naphtha, paraffine, oils; products estimated by the Irish Peat Company; heating power of peat-turf (not condensed) and peat-charcoal; relative calorific power, by the "Litharge test;" peat coke compared with coal coke; lead melted and water heated; specific gravity of peats; peat coke superior to charcoal; calorific value or power of peat coke equal to coal coke—evaporating powers of the two nearly equal; freedom from sulphur; decided superiority for working iron, brass, and copper.

TESTS, EXPERIMENTS, AND TESTIMONY 261

Statements concerning trials of peat-fuel; the practicability of using it; relative value and heating power; purposes for which it may be easily, effectually, and economically used; statements and opinions of practical and scientific men; miscellaneous tests and statements; peat-charcoal, its density and calorific power; working iron; generating steam; for domestic purposes; intensity of heating power; how to use peat; appliances for burning peat; gas from peat; gunpowder and fireworks; chemical products; analyses and properties; specific gravity of peat and peat charcoal; how to compare peat with other fuels.

PEAT FOR PAVEMENTS 263

How prepared with carbonate of lime and coal tar; resembles asphalt.

PAPER FROM PEAT 263

Experiments in France and America.

PEAT FOR BUILDING AND ORNAMENTAL WORK . . 264

How prepared; used for ornamental work on buildings, for toys, fancy articles, rings, and jewelry; resembles India-rubber.

PEAT FOR TANNING LEATHER 265

Said to possess properties valuable for this purpose.

ANTISEPTIC PROPERTIES OF PEAT 266

Remarkable cases of the preservation of human bodies, animals, &c., in peat; quotations from Dr. Rennie and Professor Lyell.

PEAT AS A DISINFECTANT AND DEODORIZING AGENT 269

Properties well known; "Chemical Deodorizing Powder" nothing but peat; "it is really the abater of every nuisance."

PEAT AS A FERTILIZER 270

Its value to agricultural interests; ingredients, qualities, and properties; analyses; weight, compositon, and comparative value.

ASHES OF PEAT 274

Value, use, and composition; used for cement and polishing powder.

CONCLUSIONS 274

The subject is of sufficient importance to command earnest attention from the business man and the philanthropist.

AUTHORITIES QUOTED 275

UTILIZATION OF COAL DUST WITH PEAT 277

A compound fuel of excellent quality, produced at moderate cost, and specially adapted for steam service; method and cost of manufacture; facilities for producing it along the lines of railroad; saving in cost of fuel; report of experiments on Western Railroad; results and conclusions; importance of this invention to mining and gas companies.

APPENDIX 287

Mainly devoted to extracts from newspapers from all sections of the country, embracing a great variety of statements and observations, of no less interest and importance than those contained in the body of the work, and equally worthy of careful perusal and consideration.

INTRODUCTION

TO THE THIRD EDITION.

THE favor with which the two former editions of this compilation of facts were received, far exceeded our anticipations.

That they have been extensively read, we have abundant evidence, from the universal interest which has suddenly been excited upon the subject throughout the whole country, more especially in the Northern States and Canada, and the demonstrations which are now being made, in numerous places, for the manufacture of peat-fuel the coming season, on an extensive scale.

That this will have an important bearing upon the prices of fuel generally, and especially upon the monopolies in the coal trade, none can question; while as a legitimate business, claiming attention through a region of country far exceeding in extent our coal-fields, it gives promise of success to a degree rarely apparent at so early a stage in any enterprise, — and for the one prominent reason above all others, that *the people* have a common interest in it.

The work is now revised and enlarged for the second

time; and extracts from numerous articles, mostly from papers published in different sections of the country, showing more clearly than could be done from any other source the widely extended interest felt in the subject, and the views entertained in all parts of the country concerning it; many of which contain valuable information of either general or local interest, are added in an Appendix.

Thus far, comparatively few experiments, of which we have any *accurate* reports, have been made with peat-fuel in this country; but the present year will doubtless witness many; and we shall esteem it a favor to be informed of such, as much in detail as practicable, with a view to incorporating them in our next edition, or giving them earlier publicity in some other form.

T. H. L.

BOSTON, March, 1867.

PEAT AS AN ARTICLE OF FUEL.

PRELIMINARY REMARKS.

"THE importance which in every age, from the earliest period of human existence, must have been attached to fire, and the necessity which has ever impelled mankind to provide for it, not so much for purposes of luxury as an absolute essential to enable them to counteract the effects of climate, and other external influences which affect the human frame, are sufficient, apart from any other considerations, to impress every one with a sense of its usefulness."

Nations, however rude or barbarous, have always made use of fire as a source of comfort or luxury, as a means of preservation, or as a destructive agent. The history of its application in relation to these three objects would go far towards illustrating a comprehensive view of the advance of civilization.

It is not in reference to these primary applications, however, that the full value of fire, or the extent of its influence, will be understood, but only when it is studied in connection with the various natural and artificial transformations of matter which it tends to produce.

Not only do the attributes of fire exert a gigantic influence in the various social requirements, but in the most minute, as well as the most elaborate, changes which take place in nature.

Heat and light, indeed, seem to be the life-giving principles of the material world, and are not less essential to man in subduing matter to his service.

Some kind of fuel has always been an article of prime necessity to man; at least from the time when he began to prepare his food by the heat of fire, or had learned to prize its comfortable warmth in the cold of winter. As experience was gained in the properties and uses of materials about him, the applications of fuel to supply his increased wants were multiplied.

By means of it, clay was converted into better bricks than those baked in the sun; limestone was burned for cement; and the ores were made to give up the valuable metals which they held concealed with a grip so fast that nought but fire could disengage or reveal them; and the subsequent treatment of these for obtaining the articles they were fitted to produce was also wholly dependent on the use of fuel.

So, from the fruits of the field were obtained, by various processes, alike dependent on the combustion of fuel, new products, the continued preparation of which, in many cases, adds not a little to their value. Indeed, most of the operations in the useful arts require, directly or indirectly, the application of artificial heat.

But the comparatively modern discovery of its being the most available source of motive-power, has given to it a new importance, hardly inferior to that derived from its other uses, causing it to contribute more than all other resources of nations to their wealth and prosperity.

It may be said that the political power of the United States, Great Britain, France, or any other civilized country, is due, not so much to their armies or navies — for these are defensive, not productive — as to the great development of their manufactures; and these are in great measure dependent upon an abundant supply of fuel, easily procurable, and at cheap rates.

The means of obtaining this, then, are of chief importance in every manufacture; and the questions of its supply, preparation, and most economical application, are of the highest interest.

Among the material sources of heat, all the substances chemically termed *combustible* may be regarded as particular kinds of fuel, although the name is usually restricted to organic products of ligneous origin, such as woody matters, peat, coal, and the like.

The substances usually employed as fuel are *wood*, *peat*, and *coal*, either in their natural state, or modified by peculiar treatment.

The abundance of all or either of these in a country must always constitute a principal source of its wealth, more especially since steam has become the moving-power of manufacturing industry, as well as the great agent in locomotion.

It is evident, therefore, that none of the productions of nature should be more carefully husbanded than those which can be used for fuel.

Every attempt also to improve the quality of inferior materials, so as to increase their efficiency as heat-producers, and consequently their value, should be liberally encouraged.

The term "fuel" is commonly applied only to substances originally derived from the growth of plants, as wood, peat, charcoal, coke, and the various kinds of mineral coal. Even thus limited, it might properly include the inflammable gases and oils, which are used of late for the sake of the heat generated by their combustion.

For objects requiring a quick heat, and at the same

time diffused over a considerable space, the most inflammable fuels are found most efficient.

The results of numerous experiments, practical as well as scientific, go to show that peat, in its rudely prepared state, goes far towards answering these requirements; and, when solidified, it is for most purposes superior.

Wood, peat, and coal, though so different in physical appearance, are, nevertheless, very closely allied in composition; all the three being chiefly composed of ligneous fibre, a compound of four simple elements — carbon, hydrogen, oxygen, nitrogen.

Physical effects have induced certain changes in some kinds of peat and coal, which cause them to differ considerably in their properties from woody fibre; but, by observing the action which analogous artificial agencies exert upon the latter, a remarkable coincidence is observed; and sufficient data are found for inferring that woody fibre is the basis of these substances; although, in the course of time, they have passed through various chemical transformations, differing in some particulars, according to locality, temperature, &c.

The analogy which exists between peat and coal is, perhaps, more readily perceived; and it may reasonably be inferred that coal, like peat, has been produced by the decomposition of species of organic growth. All who have given attention to the composition of the two substances, and the geological positions occupied by each, seem to concur in this view of the subject.

It may be said that the process which has operated to convert countless reproductions of plants into peat-bogs has been similar in the case of coals, to some extent; but, geologically considered, it is evident that the oldest peat-deposits are of modern formation, when compared to the most recent beds of coal.

Of the vegetable origin of the former no doubt can be entertained: it is apparent to the most casual observer. Of a like origin of the latter there can be no more doubt, since, among other indications, large masses of vegetable forms, and even trees, have been found in their natural positions, converted into coal; and, in the more compact varieties, the microscope has revealed conclusively a similar origin.

We do not propose to follow this portion of our subject, interesting as it is; but, leaving geologists to discuss and settle disputed points as best they may, and referring our readers to papers *ad infinitum*, which have been published upon the subject, we shall take it for granted that peat had an origin; that it exists, and is "good," and apply ourselves to practical matters concerning it.

Of peat as an article of fuel, comparatively little is really known in this country; although in numerous places in the Northern States, especially in the New England States, it is now, and has been for many years, used to a limited extent, in its crude state, and highly esteemed for domestic purposes. The writer, during a series of inquiries and investigations on the subject, has often been amused at the warm interest manifested by some honest old farmer, as he gave a glowing account of the comforts of a rousing peat fire on a stinging cold night. When properly cured, it burns freely, gives a steady and intense heat, and the uniform testimony of those who use it bears witness to its superiority in many respects.

Of wood and coal, in all their variety, the manner of preparation, and use as fuel for domestic and manufacturing purposes, the community may be said to need no information: their use is so common and universal that all have constant practical experience of their nature and value.

Not so with peat: it is by no means so generally used. Its value may be said to be unknown; and even those who have used it, in its crude state, do not appear to realize the increased value it would possess, and the extent to which it might be used, especially for manufacturing purposes, if properly prepared, and placed in the market.

We have, then, to treat of it, to some extent, as a new article; and, without in any wise attempting or pretending to offer all that might be said of it, it will be our aim to give briefly such facts in regard to it as have come under our own observation, or which we have been able to gather from a variety of sources, and that with a view to interest the community in the development and use of those rich resources of fuel which lie about us, in quantities sufficient for the demands of ages, and which require only ordinary enterprise and skill, with moderate means, to develop.

There has been comparatively little published in this country upon the subject; and the experiments made, and the results realized, have not been so scientifically conducted, nor so accurately reported, as in Europe. Much has been done in both countries to demonstrate its properties and value; and it seems to have been a common aim of all who have undertaken its manufacture or use, to condense and solidify it, and put it in merchantable shape.

The number and variety of machines and devices which have been invented, patented, or attempted to be used, for these purposes, are astonishing; and, although all have agreed that such results were practicable, few have actually arrived at anything like a satisfactory method of preparing it; and none, until recently, so far as we are aware, have arrived at that complete success which is essential to the profitable and universal introduction of an article of this character.

It is now apparent that most of the attempts referred to have failed of success from the fact that the nature of the article was not comprehended; and the principle generally started upon, to wit, that it could be condensed and produced in good merchantable shape by means of powerful pressure, applied in one form or another, was wrong. This will hardly be credited; but facts prove it to be the case.

Peat is a curious substance, possessing peculiarities of a very interesting character, some of which will appear as we progress in these pages. The fact that it is exceedingly elastic, presenting in this respect some of the characteristics of India-rubber or gutta-percha, and also that it is remarkably tenacious of water, will account to some extent for the impossibility of producing, by pressure alone, a solid, dry substance.

A process has, however, recently been discovered and applied, by which peat may be converted into a solid, dry fuel, in good shape, in large quantities, and at moderate cost. It is demonstrated, beyond a question, to be a perfect success.

The machinery is exceedingly simple in its construction and operation, and is by no means expensive when compared with the amount and value of the fuel produced by it. It is now being extensively introduced in various sections of the country. Like most inventions of the present day, it has been patented; but it is the aim of the parties having the control of the matter to encourage and stimulate the manufacture of the article; and to this end they are granting the right to work under their patents, and furnishing the necessary machinery and instructions, at rates which are within the reach of any enterprising man.

Of the purposes to which peat as a fuel can be applied, and the manner in which it can be used, the range is as wide as for wood or coal, or both.

For domestic purposes, — as the heating of dwellings, whether by furnace, or any of the innumerable varieties of stoves, or the open grate, — it is equal, if not superior, to wood or coal of any kind, save only the fact that it requires, in most cases, to be replenished more frequently than coal; but it gives a more steady, intense, yet mellow and agreeable, heat than any other fuel. In open grates, as a substitute for cannel coal, it is admirable, and produces the most cheerful fire imaginable.

In manufacturing and mechanical establishments it is available wherever fuel is required, and for many purposes possesses characteristics which render it decidedly superior; as, for instance, the production of iron and steel, and the working and manufacture of them, where the simple fact of the entire absence of sulphur, or any substance prejudicial to the quality of the metal, is a consideration of immense value.

For generating steam, it is, when solidified, second to no other fuel, and superior to most. It ignites freely, burns with considerable flame, gives an intense heat, and leaves no residuum, except a fine, light ash, which passes off freely, and leaves the grate-bars always free and clear, — a consideration which will be readily appreciated by any fireman or engineer who has had a single day's experience with the dross and clinker which is inevitable where coal is used.

A mass of facts have from time to time been published, which go to prove the truth of these statements;

and the few which we shall be able to give in these pages, and the authorities to which we shall make reference, will, we think, satisfy even the most sceptical that the subject is at least worthy of investigation and experiment.

If, then, we have at our own doors an article of fuel equal or superior to that which we now bring from a great distance, and upon which we are, and for many years have been, mainly dependent, is it not apparent that an immense field for enterprise is open to us, even though the actual gain were confined to the single item of cost of transportation saved? But it is probably true, that, in ordinary times, peat can be excavated, prepared, and cured ready for use, at less cost than coal can be mined and prepared for shipment.

Through a very large portion of the territory of the Northern States, the deposits of peat are so freely distributed that it would probably average as near a market or place of consumption as does the ordinary supply of wood now used for domestic purposes.

For the supply of iron-works, machine-shops, and manufacturing establishments, whether for the purpose of working the metals or generating steam for power, it will, in most cases, be found that deposits of peat lie within a short distance of the place of consumption; and, for some of our largest establishments, requiring immense amounts of fuel, it is known that supplies of an extent equal to their requirements for many years, lie almost at their doors.

So, too, for our railroads peat is *the* fuel; it is easily handled, ignites almost instantly, burns freely, leaving no residuum excepting light ashes, so that the grate-bars are always clean, and generates steam in a manner to charm the most exacting engineer.

There are, along the line of every railroad in New England, deposits of excellent peat, equal to their requirements for years to come; and it would seem more than probable, in view of all these facts, that, within a very short time, many of our manufacturing establishments, and all our railroads, will have each their own peat-meadow and fuel-factory; the practical results of which will be, as relates to manufacturing establishments, either a reduction in cost to the consumer of the articles and fabrics produced, or increased dividends to stockholders, or both; as relates to metals, and the various articles into which they are manufactured, superior quality, temper, &c., and a consequent increase of value, without increase of cost; and as relates to steam-power for transportation by land or water, a diminution of cost, which may inure, by the reduction of rates, to the benefit of the travelling and commercial interests, or, without reduction of rates, to the gain of stockholders; or, by such management as may be most reasonably anticipated, the results would, in all probability, prove favorable to the interests of all concerned.

The subject of peat-fuel will also be found to be of vital interest to numerous mining companies, from this fact, if from no other, that, in many cases where the yield of ores is large and rich, the scarcity, and consequent high cost, of wood or coal for smelting purposes, is such as to preclude the possibility of working them at a profit, the expense for fuel being actually, in some cases, more than the value of the metals produced.

Cases similar to the above have already come to our knowledge, where, upon examination, abundant deposits of peat have been found in close proximity to mines so situated, and which, but for this, must inevitably have been abandoned.

Prof. Emmons remarks, "There is one consideration which commends itself to the philanthropic of all our large cities; viz., the introduction of peat as a fuel to supply the necessities of the poor. It is believed that much suffering may be prevented and much comfort promoted by the use of peat in all places where fuel is expensive, as in New York and Albany. A careful examination, therefore, of places favorable to the production of this substance is a matter of some considerable importance, as it is the next best substitute for the more expensive article, coal; and anything for fuel which will save a further destruction of the forests, both in New York and the New England States, is worthy of adoption, from more considerations than one. A due proportion of woodland to that under tillage adds greatly to the beauty of any district of country; but, above all, the preservation of timber-lands is becoming a matter of great moment, and calls for legislative aid and encouragement." He adds, "We have, therefore, in this homely substance of peat, an invaluable article, of which prejudice alone can prevent a general use."

Without further preliminaries, we will proceed to give, briefly, some of the theories or probabilities as to the origin and formation of peat, the localities in which it is found, the principal varieties, some of the methods by which it has been prepared, others, perhaps, which have been attempted, some facts as to the manner in which it has been used, and the purposes for which it has been used, with such other incidental matters of scientific or practical interest pertaining to the subject as may have come to our knowledge, together with such reference to published information as will enable any so disposed to pursue the subject to an extent beyond the limited space which it is our purpose to devote to it.

Origin and Composition of Peat.

Peat is the spongy substance, found in almost every country, filling up cavities in the surface, and constituting what is termed *bog*.

It varies in color from light brown to black, and in consistency from that of a bran paste to that of clay in the bank.

The natural history of peat has puzzled inquirers a great deal; and explanations of its origin, hardly less discordant than those recorded on the subject of fossil coal, have been entertained and defended. It is unquestionably of vegetable origin, and is the result of decomposition to a certain stage, modified or affected by the agency of air, water, temperature, time, and pressure. It was once supposed that this formation was, in point of time, coeval with the disposition of the face of the country into hills and valleys. By some it was considered a bituminous deposit from the sea, — the wreck of floating islands previous to the great convulsions which the earth underwent during the formation of the present continents and islands.

By others it was even regarded as an organic substance in a state of vitality, and actually growing; but these, and many other notions and theories at times entertained, are abandoned, and more rational and philosophical views of its nature and production are adopted.

It is found, on examination, to be composed of vegetable matters, generally mosses and species of aquatic plants, in different stages of decomposition; and from this circumstance, as well as from the general appearance of the localities where peat abounds, its formation is generally accounted for somewhat as follows : —

Where pools of water collect, the soil under which is retentive, the water, not being absorbed, stagnates, and, provided the surface evaporation is not great, forms a pond. Around the borders of this pond various kinds of aquatic plants — sedges, rushes, &c. — soon make their appearance, and, by reproduction, gradually creep in towards the centre, until the whole surface becomes covered. In process of time, when several races of these have succeeded one another, and mud and slime have accumulated at the roots and around the decaying stems, a spongy mass results, which is well calculated for the propagation of mosses.

Under a constant supply of moisture, these various species thrive, continue to luxuriate, and, by progressive growth, ultimately give rise to a composition in every respect similar to that constituting the various peat-bogs.

That some such natural process has been the cause of the production of peat, appears from its composition, and the localities in which it is found. These are chiefly in the temperate zones, where evaporation is slow, and the atmosphere is generally more or less saturated with humidity.

It may be conceived, that, in the origin of these formations, the retention of the water, whether from rain or springs, in extensive basins, led at first to the development of vegetable growth in the manner above indicated; and that, the necessary moisture being supplied in abundance, the accumulation became so rapid, that ultimately the surface assumed the appearance of land; and, as decomposition proceeded, a degree of solidity was given to the mass, equal to the support of denser bodies, such as shrubby plants.

It would appear that this organic growth was rarely restricted to the original basin, but that, as it accumulated, it spread over adjacent land, which in time became a morass.

Evidence conclusive of this exists in the fact, that whole forests of almost every description, such as oaks, firs, ash, birch, yew, willow, &c., have been overwhelmed in its gradual but steady advancement, and are found in all positions at the bottom of peat-bogs.

Generally this formation is met with in climates of a moist nature, in level countries, where imperfect natural drainage exists; although it is found in considerable beds in upland districts.

In mountainous districts, in addition to the imperviousness of the rock to the moisture, the constant formation of clouds upon those elevated regions favors the growth of the mosses and plants, the decomposition of which contributes to the increase annually of these deposits.

In America, peat is rarely found in these elevated positions, and then only in small quantities: but in Great Britain, and on the Continent, the deposits are numerous and extensive; and, as a general thing, they are esteemed of superior quality for fuel. Instances of this kind are frequent in Ireland, Scotland, Northern Germany, and Holland, while others are found high up the Alps, in the Vosges and in the Jura.

The extent and depth of the peat-bogs vary considerably in the different countries where they are found, and seem to depend upon circumstances quite distinct from each other.

It is evident that the area which they may occupy is intimately connected with the distribution of water.

Their depth in this country varies, generally, so far as we are able to learn, from one to twenty feet: though, in some instances, it is reported to be twenty, thirty, fifty, and even eighty feet; but an average of the depth of what may be considered our peat regions would probably be not far from five or six feet.

The morasses of Holland and Germany are, to a considerable extent, about six feet deep, as are likewise those in upland situations; while many of the peat-bogs of Ireland are from thirty to forty feet in depth; and peat-banks of solid black fuel are exposed in some localities to a corresponding height.

On intersecting these deposits, whatever their thickness, it appears, from the fact of layers of gravel, clay, shells, &c., being interposed horizontally, that, in many cases, these tracts have been swept over with currents of water more or less violent: such layers, however, are never more than a few feet in thickness, and seem to have retained all the conditions favorable for the continued growth of the plants conducive to the formation of peat.

From its physical constitution, this substance may be regarded as a kind of fossil fuel; and undoubtedly it is one of the most extensive sources of fuel known.

A writer in the "New American Cyclopædia" says, "The dense, compact peat appears to represent the first step in the progressive changes from vegetable substances to mineral coal."

Dana says, "Peat is sometimes entirely converted into coal."

Sir James Hall says, "I have always looked upon

the peat of the Old World as one of the principal sources of our coal."

Dr. Macculloch, from his researches, determined that peat is intermediate between simple vegetable matter and lignite; the conversion of peat to lignite being gradual, and being brought about in a great lapse of time by the prolonged action of water.

Many bodies are detected in peat, however, which are not contained in coal; although the ultimate elements of both are the same.

Viewing it as the product of the decomposition of plants, carried on through a long succession of ages up to the present time, it is natural to expect, that, when cut vertically, differences should appear.

Mr. Williams remarks, "I have seen strata of coal that bore all imaginable marks of being composed of wood: the color, the quality, the stratification, the manner of burning, the ashes, and everything else, looked like peat."

Dana says, "In temperate climates, it is due mainly to the growth of mosses of the genus *sphagnum*. This plant forms a loose turf, and has the peculiar property of dying at the extremity of the roots below, while it continually grows and increases above the surface; and by this process a bed of great thickness is gradually formed."

Dr. Rennie says, "No living animal exists in peat;" showing the advanced state of decomposition.

In almost every instance, this progressive change is exhibited; and, consequently, peat is classed by some authorities into *recent peat* and *older peat*, from the appearance it presents.

The former bears distinctive traces of its origin in the

roots, leaves, and stems of plants, the structure of which is still retained.

It is very porous, tough, and elastic in some tracts, but in others, especially where the bog is well drained, very brittle.

The color varies, with the character of the vegetable growth, the age, and the progress of the decomposition, from a light brown to black.

The latter or "older peat," to which the preceding gradually inclines, shows few traces of fibrous matters, such as roots, stems, or leaves; but it presents, when cut, a pitchy, shining hue, and is dense and fine in the grain.

Preference has always been accorded to this as a fuel, from its superior gravity, and the greater heat which it produces when undergoing combustion.

From the change which the vegetable substances pass through, it is evident that the usual process of putrefaction is carried on in the ordinary way, at the commencement; but as the surface grows, and contact with the air is cut off, the mass is left to the play of the affinity of its elements, rendered more active by the steadily increasing pressure which it has to sustain.

The occurrence of peat is frequently indicated by the growth of dwarfish evergreens and rank swamp herbage, and by the elasticity of the crust which supports them.

Dr. Rennie remarks, "Though all peat-moss be of vegetable origin, yet the situations in which it is formed, the plants of which it is composed, and the state in which it is found, being different, it is reasonable to expect that one moss should differ from another in its appearance, qualities, and the uses to which it may be made subservient. This difference may be detected by the naked eye, whether the moss be in the pit, or dug and dried, or burning, or reduced to ashes. The various colors that substance assumes, and the external appearance of it, mark the difference. Some are of a bright yellow color; others brown, or jet black: some are composed of a congeries of vegetables in an organized state; in others, few or no traces of organization can be seen. Clay, sand, and shells may be detected in some; in others, no such mixture can be discovered. Some are soft and greasy like butter, and form a hard, brittle, tenacious peat, almost like coal; others are loose and friable like mould. The water squeezed out of one moss is of the color of amber; of another, of claret or port wine; and of a third, as black as ink. In some cases this water effervesces with chalk; in others, not. Sometimes it leaves a copious sediment by evaporation, which is highly inflammable; in other cases the sediment is small, and scarcely inflammable. Some are covered with a rich luxuriance of aquatic plants; others are utterly bare, barren, and destitute of vegetables on their surface."

Peat-meadows may be easily probed and examined by means of a pole, rod, or tube, thrust through the surface, to which the peat, if present, will adhere in quantities sufficient to determine something of the character of it. A simple instrument, admirably adapted to such examinations, may be described as follows: —

Two pointed half-cylinders, of any metal sufficiently strong for the purpose, say twelve inches in length by about one or one and a half at their greatest diameter; an exterior and an interior, so arranged, that, when put together, the exterior rolls upon the interior, causing the whole to present the appearance of a conical or tapering cylinder, attached to a handle made of strong wood in lengths convenient to carry, and so arranged as to be connected at pleasure by means of proper joints. The instrument being closed, it is forced into the peat to the depth at which the examination is to be made; then, by turning the handle half round, the pressure on the exterior cylinder forces it behind the interior; and in this manner the instrument is opened. It is then forced downward about the length of the cylinder, which immediately becomes filled with the soft substance; the handle is again turned half round, which closes it, and secures the contents, which may be withdrawn and examined at pleasure.

The density of peat varies with the position in which it is found; with the organic substances from which, in different localities, it had its origin; the character and temperature of the atmosphere and climate; the proportion of earthy and mineral matters which it contains; and the thickness of the strata.

Freshly cut, it is found to be saturated with water to the extent of from thirty to ninety per cent., according to locality and density; and even when subjected to the ordinary process of air-drying, it will be found to have retained a considerable percentage of moisture for a long time, even after it has the appearance of being perfectly dry; and it will readily be understood, that, beyond a very small percentage, the amount of moisture so retained will tend to diminish very considerably the heating power of the fuel.

The degree of decomposition, however, which the substance has undergone, is supposed to determine, principally, the difference of specific weight in peat from the same cutting.

So highly inflammable are some of the denser kinds of peat, that the characteristic distinctions of bituminized wood are considered insufficient to explain the circumstance; and hence the Ince peat of Lancashire is believed by some to be penetrated by petroleum from bituminous springs.

Instances of similar character have been reported in this country, but without sufficient evidence of the fact to render it certain that such is the case.

In its natural state, peat is highly antiseptic; to which may be attributed the long period which is required to elapse before the traces of the substances originating it are effaced.

This quality has been further proved by the wood which is found without being in the least injured, although buried in it to great depths for centuries. Not only these, but human bodies, and remains of animals, — the latter, in some cases, of extinct species, — and other substances prone to putrefaction, have been found at great depths in a high state of preservation.

In the Philosophical Transactions for 1757, mention is made of "a stratum of peat on each side of the Kennet, near Newbury, in Berks, which is from a quarter to a half mile in width, and many miles in length. The depth of peat is from one to eight feet; and great numbers of entire trees are found lying irregularly in it. These are chiefly oaks, alders, willows, and firs, and appear to have been torn up by the roots. Many horses' heads, and bones of several kinds of deer, the horns of the antelope, the heads and tusks of boars, and the heads of beavers, are also found embedded in it.

Peat always contains some earthy matters in greater or less proportion, according to the thickness of the stratum, and its position relative to soil in the surrounding region.

Surface peat generally contains less of mineral matter than the second and lower strata, which sometimes possess so much of these as to render it useless, in point of economy, as fuel.

In this, as well as from the nature of those matters, it differs essentially from wood.

These substances are left, when the peat is consumed by open combustion, in the form of ash; and, from the nature of the ingredients, it presents various appearances, from white to gray and ochrey-red.

The percentage of ash differs widely, as may be supposed; being reported in the statements of numerous results, to which we have had access, all the way from one to thirty-three per cent. The better qualities, however, or those which may be said to be most used, would seem to yield a quantity ranging from three to eight per cent.

The varieties yielding the largest amount of ashes are valuable as fertilizers, from the fact that they contain a large amount of phosphates, and other salts which serve to enrich the soil.

Copious details of the composition of the ashes of many of the peats of Ireland are given by Muspratt, in his "Chemistry as applied to Arts and Manufactures."

Referring to the odor emitted by burning peat, Dr Macculloch says, "It is occasioned by an essential oil. That odor may, therefore, be useful for the lungs and head. Indeed, there is a traditional opinion among the Irish people, that those who use peat-fires are less liable to consumption than others."

Mr. Jameson, in his "Mineralogy of the Scottish Isles," remarks that peat is peculiar to cold or cool climates; and thus Nature has provided the means of a constant supply, from this source, of the necessities of the people who dwell in those regions, and who continually require fuel.

As we advance towards the warmer climates, vegetable matter is more rapidly decomposed; until, at the tropical regions, the putrefaction of animal and vegetable matters is so rapid, that it prevents the formation of any body of the substance and structure of peat.

In Scotland, it is observed that the peat at the bottom of a mountain is more decomposed than that which occurs at its top, and that the lignites found in turf mosses or bogs are more sound upon the summit of a mountain than at its base.

It is also observed that the peat of the south of England is more decomposed than that of the north of Scotland, and the peat of France has more of the coaly appearance than that of England.

Prof. Lyell says, "It has seldom, if ever, been discovered within the tropics; and it rarely occurs in the valleys, even in the south of France and Spain. It abounds more and more in proportion as we advance farther from the equator, and becomes not only more frequent but more inflammable in northern latitudes; the cause of which may probably be, that the carbon and hydrogen, which are the most inflammable parts, do not readily assume the gaseous form in a cold atmosphere."

Darwin states, that, in the southern hemisphere, peat does not occur nearer to the equator than latitude 45°; that the composition there met with, results from the decomposition of the plants and grasses. The fact that no mosses, so far as can be ascertained by strict examination, enter into the species of peat found in South America, favors this view.

It would be difficult to say at what period this material

was first applied to its long-acknowledged useful purposes. That it was used, as it is at present, from a very early period of our history, there can be no doubt; and, in the absence of ligneous and mineral fuels especially, its great abundance, the comparative ease with which it is obtained, and singular production, arrest attention to one of those sources of comfort and convenience which an infinitely wise Creator has laid up in the storehouse of Nature for the benefit of mankind. It is clear that its value for purposes of fuel was early understood in Germany; for Pliny says expressly, that the Chaucé pressed together with their hands a kind of mossy earth, which they dried by the wind rather than by the sun; and which they used not only for cooking their food, but also for warming their bodies.

Beckmann, in his "History of Inventions," mentions a letter of sanction, by which an Abbot Ludolph, in the year 1113, permitted a nunnery near Utrecht to dig *cespites* (turf) for its own use in a part of his *vena* (turf-bog). On the same authority, we are told that the words *turba*, *turbo*, *turbæ*, and *turfa*, occur for turf in the years 1190, 1191, 1201, and 1210. The traffic in this kind of fuel is recognized in the *Leges Burgorum* of Scotland as early as about 1140. *Turbaria*, for a turf-moor, is found in the writings of Matthew Paris, who died in 1259. *Turbagium* is found in a diploma of Philip the Fair, in 1308, where its connection is such as to signify the right of digging turf. Brito, who lived about 1223, is quoted as mentioning turf among the productions of Flanders. It may be added, that words of like signification occur frequently in the earlier foundation charters of the monasteries in Germany, as conveying the right to dig turf generally within a certain limited extent of ground.

Pliny (Hist. Nat. Lib. XVII.) expresses his pity for the "miserable people" living in East Friesland and vicinity, in his day, who "dug out with the hands a moor-earth, which, when dried, they used for preparing their food and warming their bodies."

Charred peat is said to have been used in the Freyberg smelting-houses about the year 1560; and mention is made of its use for like purposes in England in the early part of the seventeenth century.

Dr. King, an Irish writer, in 1685, says of turf, "It is accounted a tolerably sweet fire; and having very impolitically destroyed our wood, and not as yet found stone-coal, except in a few places, we could hardly live without some bogs. When the turf is charred, it serves to work iron, and even to make it in a bloomery or iron work. Turf charred I reckon the sweetest and wholesomest fire that can be; fitter for a chamber and for consumptive people than either wood, stone-coal, or charcoal."

Methods of Preparation for Fuel.

Where peat-bogs abound, and the inhabitants make use of it as fuel for domestic purposes, the process of preparation is very simple, and has varied little, if any, for ages.

The surface-layer, or turf, which contains the living plants and their roots in the natural state, is stripped off to the depth of six, nine, or twelve inches.

The material is then cut with a kind of spade known as the *slane*, which has a blade about fifteen inches long by four and a half inches broad, with a wing on the side, bent upwards at right angles to the blade, so as to form, with the latter, two sides of a square.

With this the peat is cut in long, square masses, and laid upon the sward, where it spontaneously loses its water, partly by infiltration into the soil, and partly by evaporation.

The tool is worked by the arms alone, and each block is cut by a single motion.

After the blocks are partially dry, having been turned at intervals so as to expose the different sides to the sun and air, they are found to be reduced very materially both in size and weight, and to have acquired a good degree of consistency. They are then piled or cobbled up in heaps on the sward, care being taken to dispose them in such manner as will admit of a free circulation of air through the mass; and, after remaining exposed in this manner for some weeks, they are generally removed to some airy place of shelter, where the process of drying may continue, and the fuel be convenient of access when the season arrives for its consumption.

Such is the mode generally adopted, both in this and other countries, when the peat is of sufficient density and elasticity to bear being so handled without breaking.

When, however, the material is brittle, and will not admit of being used in this way, it is dug out with ordinary spades and shovels, and all roots, sticks, stones, and such like bodies, picked out. It is then spread upon the greensward, or, in some cases, upon suitable ground covered with a layer of straw or hay, in a mass, to the depth of eight to eighteen inches, with a breadth of about four or five feet, and to such lengths as may suit the convenience of the laborers. In this condition it is brought to a homogeneous mixture by harrowing, raking, working over with hoes, spades, or other tools, or by the treading of men or animals, until it is of about the consistency of stiff mortar, when the surface and sides are smoothed, and it is left in this state to drain and dry.

After remaining for one, two, or three days, according to the weather, and acquiring a somewhat greater degree of consistency, it is rendered still more compact by beating the surface with shovels, spades, or paddles adapted for the purpose; and in some parts of Europe this is accomplished by treading, which is there mostly done by women and children, who attach flat boards, about six inches broad, and twelve to fourteen inches long, to their feet.

By this time the peat has acquired such solidity that it will bear a person's weight upon it without sinking.

The surface is then marked off, or cut by the sharp edge of a board, or a large knife adapted for the purpose, to the depth of one or two inches, into squares; the sides of which are from three to six inches, according to the size desired for the fuel when it shall have been thoroughly dried, and ready for use.

In this condition it is left to dry; and, as evaporation proceeds, the squares contract, the cuttings gradually open down to the bottom, and the mass is separated into blocks of somewhat uniform size, standing on end, and of pyramidal form; the base being still quite moist, and covering nearly the whole surface, while the top, which has been most exposed to air and sun, has contracted to nearly or quite one quarter of its original size, and is dry and hard. The blocks are then turned once or twice, in order to give a more uniform exposure; and at the expiration of a few days of good weather, they are in condition to be removed, and stored for use; care being taken, however, that it be in a sheltered but airy location, and that it be not too closely packed; for, notwithstanding it may have the appearance of being quite dry, it will be found to have still retained a very considerable amount of water, and, if too closely packed, is liable to a fermentative process, which injures the quality, and has been known to raise the temperature so high as to cause spontaneous combustion.

It is by a process quite similar to this, in most respects, that peat has recently been manufactured at Barnstable, Mass., which was sold in considerable quantities in Boston during the winter of 1864–5, and met with very general favor.

It was sold in large quantities at $8.50 per ton, and retailed at $9.00 per chaldron of thirty-six bushels, $1.00 per barrel, and 40 cents per bushel.

Experiments, to a limited extent, were also made with it in Boston and vicinity under stationary and locomotive boilers, with highly satisfactory results. It cannot, however, be recommended for general use for these purposes, unless solidified to a much greater extent than by this method.

Another simple process, quite common in Holland and Ireland, is to take of the mass, after being reduced to a uniform consistency, somewhat like stiff mortar, as before described, in bulk equal to the size of an ordinary brick, and by a rapid process of manipulation with the hands, easily acquired and practised, to form it into oblong blocks of somewhat uniform size and shape, which are laid upon the ground until such time as they have acquired a sufficient degree of consistency to be "cobbed" up, or stored for further curing and use, as in the process last before mentioned.

This is called "hand peat," and is considered somewhat preferable to either of the other kinds described.

Two machines for *cutting peat* are thus described by Prof. Johnson: —

"In North Prussia, the Peat-cutting Machine of Brosowsky is extensively employed. It consists of a cutter, made like the four sides of a box, but with oblique edges, which by its own weight, and by means of a crank and rack-work, operated by men, is forced down into the peat to a depth that may reach twenty feet. It can cut only at the edge of a ditch or excavation, and when it has penetrated sufficiently, a spade-like blade is driven under the cutter by means of levers, and thus a mass is loosened, having a vertical length of ten feet or more, and whose other dimensions are about twenty-four by twenty-eight inches. This is lifted by reversing the crank motion, and is then cut up by the spade into blocks of fourteen inches long by six wide and five deep. Each parallelopipedon of peat, cut to a depth of ten feet, makes one hundred and forty-four sods, and this number can be cut in less than ten minutes. Four hands will cut and lay out to dry twelve thousand to fourteen thousand peats daily, or thirty-one hundred cubic feet. One great advantage of this machine consists in the circumstance that it can be used to raise peat from below the surface of water, rendering drainage in many cases unnecessary. Independently of this, it appears to be highly labor saving, since thirteen thousand machines were put to use in Mecklenburg and Pomerania in about five years from its introduction. The Mecklenburg moors are now traversed by canals, cut by this machine, which are used for the transportation of the peat to market.

"Lepreux in Paris, has invented a similar but more complicated machine, which is said to be very effective in its operation. According to Hervé Mangon, this machine, when worked by two men, raises and cuts forty thousand peats daily, of which seven make one cubic foot, equal to five thousand six hundred cubic feet. The saving in expense by using this machine is said to be seventy per cent. when the peat to be raised is under water."

He also describes a method of dredging peat, as follows: —

"When peat exists, not as a coherent more or less fibrous mass, but as a paste or mud, saturated with water, it cannot be raised and formed by the methods above described.

"In such cases the peat is dredged from the bottom of the bog by means of an iron scoop, like a pail with sharp upper edges, which is fastened to a long handle. The bottom is made of coarse sacking, so that the water may run off. Sometimes a stout ring of iron, with a bag attached, is employed in the same way. The fine peat is emptied from the dredge upon the ground, where it remains until the water has been absorbed or has evaporated so far as to leave the mass somewhat firm and plastic. In the mean time a drying bed is prepared by smoothing, and, if needful, stamping a sufficient space of ground, and enclosing it in boards fourteen inches wide, set on edge. Into this bed the partially dried peat is thrown, and, as it cracks on the surface by drying, it is compressed by blows with a heavy mallet or flail, or by treading it with flat boards, attached to the feet, somewhat like snow shoes. By this treatment the mass is reduced to a continuous sheet of less than one half its first thickness, and becomes so firm that a man's step gives little impression in it. The boards are now removed, and it is cut into blocks by means of a very thin, sharp spade. Every other block being lifted out and placed crosswise upon those remaining, air is admitted to the whole, and the drying goes on rapidly. This kind of peat is usually of excellent quality. In North Germany it is called '*Baggertorf*,' i. e. mud peat."

Several peat-cutters have been devised, though we have seen no published description of any except those referred to above.

Our own experience, confirmed by the testimony of others, and by the observations of those who have had opportunities to see and investigate where we have not, is, that for cutting peat in this country, there is no better machine than the simple *slane*, in the hands of a stout, good-natured Irishman, well treated and fairly paid.

Such a "machine" will easily cut fifty tons crude peat per day of ten hours. This statement is often questioned; but the following simple facts and figures will be found to demonstrate it, and leave a margin besides.

The blocks of peat, as ordinarily cut with the slane, from a well-drained meadow, weigh from twelve to twenty-five pounds — say fifteen pounds, which is less than the average. A man *can* cut twenty-five or more blocks in a minute; when working "by the job" he *will* cut twenty to twenty-two; and when working "by the day," he *does* cut fifteen or more. If, then, taking these lowest figures, say fifteen blocks of fifteen pounds each, we have a result, two hundred and twenty-five pounds per minute, or six and one fourth tons per hour, and consequently sixty-two and a half tons per day of ten hours.

In order to work a bog which is entirely submerged with water, or which is saturated with water to such extent that it will not retain its form when cut out in the ordinary manner, it is evident that dredging must be resorted to.

A variety of machines for this purpose are in use,

some of which we have seen in operation for harbor, canal, and river service, but not in a peat-bog; and our own experience has not been such as to enable us to designate the peculiarities of structure requisite to determine which would be the *best* for dredging peat.

In the "Journal of the Society of Arts," for November, 1862, was published a paper on "The Utilization of Peat," by B. H. Paul, from which we extract a few paragraphs: —

"In one case we find peat deposits in the form of what are called peat-bogs, — masses of peat of considerable superficial extent, and generally of great depth, — twenty, thirty, and sometimes upwards of a hundred feet deep.

"In the other case, we find, situated on the slopes of mountainous country, peat deposits which are never of very great depth, generally from two to twelve feet, and where the peat is sufficiently solid to be walked upon with ease. In these deposits the peat is of a more uniform texture and character throughout than in bogs, although there is always a greater or less difference between the peat at the surface and that at the bottom. These deposits of mountain-peat are very common in the Highlands of Scotland and in some parts of Ireland. Mountain-peat offers very much greater facilities for cutting than bog-peat, and it is generally of much better quality; capable of taking a high polish when rubbed, and of a density greater than that of water, the cubic foot weighing from fifty-three to seventy-eight pounds.

"The method of cutting peat in the Highlands of Scotland is very different from that adopted for cutting from bogs. After removing the surface-sod, the peat-cutter, with a peculiar shaped tool, digs out the peat in slices of about a foot square, and three or four inches thick."

The process of cutting and curing is given at some length, and he proceeds: "When mountain-peat is cut in slices, as I have described, and spread out on the ground during dry weather, the drying goes on rapidly; the surface of the pieces acquires a kind of skin, which is not wetted again by rain; and the peat in the course of a week is sufficiently hardened to be handled. The pieces are then set up on edge, so that the air may play on both sides; and, in the course of six weeks or two months, they are dry enough to be stacked or heaped up. In the Highlands of Scotland and in the Hebrides, on the average, there is rain four days out of six; and it is only during the months of May, June, and July, that any continuance of weather favorable for drying peat can be expected. It is necessary, therefore, to obtain the utmost advantage of that period for drying; and to do so the peat must all be cut before the end of May, at latest. On the other hand, if the peat is cut during frosty weather, and becomes frozen, it crumbles to powder when the thaw comes; and for this reason it is not safe to commence the cutting at all before April, or even May. As a rule, it may be said that the month of May is the only time available for cutting peat in the Highlands of Scotland, and more especially in the Hebrides, so as, on the one hand, to avoid the destruction of the peat by frost, and, on the other hand, to insure the best possible chance of getting it dried.

"Two men working together, one cutting and one casting the peat, will, in good weather, get through what is equivalent to ten tons of dry peat; so that, if they were able to work every day during May, they would cut from two hundred to three hundred tons of peat; and, to get ten thousand tons cut and spread, one hundred men would be required for the whole month."

One essential quality of peat in relation to its value as a fuel is its density; and consequently numerous efforts have been made, and various processes have been attempted, by which to give it a degree of solidity equal or approximating to hard coal, and sufficient to stand the blast required for a very high degree of heat in the more severe processes of metal manufacture and steam service.

This has generally been attempted by means of direct pressure, variously applied, upon the raw material as taken from the bogs. Radical difficulties, however, stand in the way of effecting the desired condensation by any such methods, owing principally to the elastic nature of the article itself, increased to a considerable extent in all peat of a fibrous character, which causes a distention after the pressure is removed.

None of these methods of *pressure* alone, have proved practically and economically successful.

We can barely allude to a few of the numerous methods by which it has been attempted to condense or solidify peat, as any extended descriptions of the various machines and processes would require space far beyond our limits.

Mr. C. M. Williams, of the Cappoge Works, a gentleman who has contributed much to the development of the industrial uses of peat in Ireland, used powerful hydraulic presses, and was able to produce a fair article in considerable quantities. It is questionable, however, if, in this particular department of its manufacture, his endeavors have been remunerative.

We shall have occasion to refer to him again as we proceed with our subject.

The manner of compressing, as conducted in the Cappoge Works, was to break up the fibre of peat as much as possible, and then to place it, interlaid with coarse cloths or cocoa matting, under the pressing machine. After remaining under pressure a sufficient time, the material was found to be reduced to one third its original volume, and to have lost about two fifths of its weight.

Mr. Cobbold's method is to convert the peat into a pulp, with the addition of water if necessary; after which, by means of centrifugal power applied to the mass, the water is got rid of, and a very dense product is obtained.

The course recommended by Mr. W. B. Stones, as described in his patent of March, 1850, is the use of a box divided into a number of compartments suited to the size of the machine, and pressure to be exerted upon this by rollers adapted to one another by means of leverages.

His claims covered a machine for pressing peat; a process of carbonizing; the application of carbonic-acid gas to the extinction of glowing char-peat; the employment of peat-gas produced during the operation of carbonizing, for the purpose of heating the retorts; the application of a series of receivers to the distillation of the residuum, and the obtaining products therefrom; a process of obtaining "peatole" and "peupion" by rectification; a process of obtaining "peatine;" the application of sulphur and peat to the manufacture of bisulphuret of carbon, and application of the peat and sulphur residuum to the manufacture of gunpowder; the manufacture of artificial fuel from anthracite and char-peat; the impregnation of surface-peat with resin-oil, &c., for the manufacture of fire-lighters and revivers; the purification of peat-gas, as described; the obtaining of heat and light by the combustion of peat-gas in atmospheric air, when a cob or plate of platinum is employed; a peculiar construction of gas-burner, and application of these burners, for the purpose of blow-pipes, &c.

At the International Exhibition of 1862, a machine was exhibited, which in the official catalogue is described as Brunton's; but it was actually the invention of Mr. Buckland, formerly of the Maesteg Iron Works, South Wales.

It consists of a solid obtuse iron cone, having a spiral groove on its exterior, and revolving vertically within a hollow cone of iron plate, perforated everywhere with small round holes just like a colander, which in fact it is. The peat is put into the space between the solid and hollow cones, and, by the revolution of the former, is forced in worm-like form through the holes in the latter.

Thus prepared, it is fashioned into bricks by any convenient machine, of which one was shown at the same place.

The bricks are artificially dried; and portions of them which were exhibited were solid and resisting.

Peat-charcoal prepared from them was also shown.

The machine is an ingenious one, and possesses some points of interest; but we do not think it can be made practically useful as a mode of manufacture. It has run several "experimental trips" in this country; and we have had repeated opportunities to see it in operation, and examine it thoroughly. The work is well done, so far as it goes. It appears to us, however, to require to be run with great care, is liable to get out of order, can do its work well only at a low rate of speed, and is equal to the production of but a small amount compared with the power required to run it, and can therefore be used satisfactorily only by parties who may desire to run it "regardless of expense."

Another method, practised to some extent in Southern Bavaria, is, after drying peat in the ordinary manner, to pulverize it by passing it through rollers; then to drive off the remaining water by heat, and consolidate the dry powder by powerful pressure.

Mr. C. Hodgson, in a paper read before the Institution of Civil Engineers of Ireland, refers to the same matter, and describes a process whereby he produces from dry powdered peat, an excellent article of solid fuel. He employs for its compression an engine patented by himself, which he describes as a horizontal reciprocating ram, working in a cylinder five feet long, with a uniform bore. The powdered peat falls into this as the ram draws back at each stroke; and soon, filling the whole length, considerable friction takes place against the sides of the tube, before the frictional resistance of the column is overcome, and the whole mass moves on; so that the blocks formed at one end are successively discharged at the other, at the rate of sixty a minute, making in an hour about fifteen cwt. of compressed peat, equal in density to coal. This apparatus was recently in operation at Derrylea, near Monasterevan; and it is said by the inventor to leave no doubt of the practicability of producing dry compressed peat on a large scale, and with profit.

Messrs. Gwynne & Co. have taken out several patents for the preparation of peat-fuel. One of their processes bears some resemblance to the method last described. The peat, as dug from the bog, is deprived of much of its moisture by being placed in a large centrifugal machine; after which it is ground to powder, and passed through a series of cylinders revolving in a heated chamber, where the remaining moisture is got rid of, and the powder raised to the proper temperature for compression. From the last cylinder it is carried by pockets to the compressing tables; and, having passed through them, the solidified peat is ready for use. It is found that when the peat-powder has been dried at a temperature of about 180°, and in that state allowed to enter the hopper of the compressing engine, the tarry properties of the turf are just sufficiently developed to form a good cementing compound; and the brick of compressed turf, when cold, forms a dense and very pure fuel.

In Austria, as well as in some parts of Germany, a process prevails, to some extent, of grinding the peat as rags are ground in a paper-mill, reduced, by the addition of a great deal of water, to a very fine and soft pulp, which is then placed in tanks, basins, or other suitable receptacles, where, by filtration and evaporation, it is, after a considerable time, relieved of most of the water, and, when of the consistency of cheese, can be cut or broken up and stacked for use at a still later day.

The product is excellent, but the process is slow, tedious, and expensive, and by no means adapted for general use.

At Horwich, in Lancashire, England, works have been erected which are reported as being operated successfully, and are said to be well conducted, and to produce an excellent article of fuel.

The peat, as it comes from the bog, is thrown into a mill arranged for the purpose, by which it is reduced to a homogeneous, pulpy consistency. The pulp is then conveyed, by means of an endless band, to the moulding machine, in which, while it travels, it is formed into a slab, and cut into blocks of any required size. The blocks are delivered by a self-acting process on a band, which conveys them into the drying chamber, through which they travel forwards and backwards on a series of endless bands at a fixed rate of speed, exposed all the time to the action of a current of heated air. The travelling bands are so arranged that the blocks of peat are delivered from one to the other consecutively, and are by the same movement turned over in order to expose fresh surfaces, at regular intervals, to the action of the drying currents; so that they emerge from the chamber dry, hard, and dense.

The next stage in the process is the treatment of the peat in close ovens, when it may either be converted into charcoal for smelting purposes, or may be only partially charred for use as fuel for generating steam, or in the puddling furnace.

In a paper recently read before the British Association by D. K. Clark, C. E., a somewhat detailed account of these works was given, extracts from which will be found in the Appendix to this work.

Prof. Johnson, in his narrative of peat operations in Europe, describes the *Mannhardt* and *Neustadt* methods, by which the crude peat in its moist state is passed between rollers, being reduced somewhat in bulk, and delivered in soft blocks or sheets, after which it is removed to the spreading-ground or sheds to be dried; also the *Lithuanian*, *Exter's*, and *Elsberg's*, which in their main features correspond with that of Gwynne, described above; *Versmann's*, which is almost identical in principle and construction with Buckland's, already described, on page 43, and *Challeton's*, *Siemens'*, *Weber's*, *Gysser's*, and *Schlickeysen's*, which, by methods similar in principle, but varying somewhat in the detail of construction and operation, pulp the crude peat, in some cases adding to it a considerable amount of water, and deliver it in moist blocks, to be dried either in kilns or in the open air.

The advantages claimed for Hodgson's, Gwynne's, Exter's, Elsberg's, and the Lithuanian methods, are, that the expensive transportation and handling of fresh peat, containing a large amount of water, is avoided, and that enough peat may be air-dried and stored during summer weather to supply a machine with work during the whole year.

It is evident, however, that the "fresh peat containing a large amount of water" must be handled and dried at *some* stage of the operations; and experience has demonstrated beyond a question that a thousand tons of crude peat, as it lies in the bog, can be taken out, manufactured, and dried *after* it has been manufactured, in less time and at less cost for labor than the same amount can be taken out, dried *before* being manufactured, and then formed into blocks, as practised by this method; yielding, moreover, a fuel of superior character in its mechanical composition.

Of this method of dry pressing, Prof. Johnson remarks, "Its disadvantages are, that it requires a large outlay of capital and great expenditure of mechanical force. Its product is, moreover, not adapted for coking. When wet, the surface of the cakes swells up, and exfoliates as far as the water has penetrated. In the fire a similar breaking away of the surface takes place, and when coked the coal is but moderately coherent.

"The idea that heat develops bituminous matter (in the process of manufacture), or fuses the resins which exist in peat, and that these cement the particles, does not harmonize with the fact that the peat thus condensed flakes to pieces by a short immersion in water."

A correspondent from Austria writes as follows: —

"Referring to the preparation of rough peat, there are five kinds in Germany and Austria.

"1. The fibrous peat is picked out and dried in the air.

"2. The earthy peat is digged out, stricken to tiles, and dried in the air.

"3. The peat is ground and put into a great deal of water, in which the peat forms by itself a very strong cake.

"4. The peat is handled like by your manner.

"5. The peat is ploughed like a field, then heaped up and carried on railroads to the factory. It is ground like rags; then it runs through a spiral, which is heated on the outside by steam or smoke of a far fire for a boiler; the hot peat falls immediately into the press, where it is pressed to very strong pieces, about seven inches long, three inches wide, and only half an inch thick. This manner patented for Mr. Exter in Munich."

These will, perhaps, serve to illustrate the general characteristics of an infinite number of methods proposed and machines devised for improving the quality of peat fuel, so as to remove or overcome its characteristic fault; namely, want of density; and its treatment for other products. We shall have occasion to refer to these again, and perhaps to mention others as we proceed.

In this country comparatively little has been done; and until quite recently no machinery whatever, specially constructed and adapted for the production of solid fuel from peat, has been put in practical operation. The impression has seemed to prevail, that the material is to be treated like clay, and that brick-machines might be readily made to work the desired results; but the idea is erroneous. Numerous brick-machines have been tried, some of them very ingeniously and perfectly constructed, and which have been demonstrated to be almost perfect in their operation upon clay, but have proved an entire failure when peat was substituted instead. It is true, however, that, with two or three of these machines, peat has been pressed into compact blocks, having the appearance of great solidity *when moist;* but so soon as the moisture is evaporated, as it will inevitably be in time, the mass is found to be porous and light.

Quite a number of presses, some of them exceedingly ingenious in device and construction, and powerful in their operation, and supposed to be so aranged as to press the water out of the mass, and leave the material compact and nearly or quite dry, have been built and tested, with failure of success as a uniform result; and although the records and reports of such cases, both in Europe and this country, are sufficiently extensive to explode the idea that any profitable results can be obtained by pressure alone, there are, nevertheless, those who are still persistent in their efforts to accomplish it by such means, and are now devising new methods of applying powerful pressure, which, were they to consider but for a moment the nature of the material in its crude state, would be seen at once to be clearly of no avail. The famous Beater Press, which within a few years has acquired great notoriety, and is probably the most powerful press now in use for hay, straw, cotton, tobacco, &c., has been tried several times in New York and Massachusetts, by parties sanguine of success, but with only the same results as with other presses.

A multitude of experiments have been tried, and all sorts of machines devised, but, in most cases, by parties who, it would seem, have failed to comprehend the

4

nature of the material, have consequently gone to work in the wrong way to obtain results, and, as a matter of course, have, in the majority of cases, been unsuccessful.

So much interest has been had in the matter within the past year or two, that several parties in different sections of this country, each in their own way, have been at considerable expense to develop it; but in nearly all of the cases, we hear, thus far, only of larger expenditures without the desired results in product or profit. We hear of these experiments on Long Island; at Nyat, near Providence, R. I.; at Schenectady, Poughkeepsie, Syracuse, Oswego, Rochester, and Pekin, N. Y.; at Belleville, N. J.; and Springfield and Worcester, Mass. In some of these cases, brick-machines have been used, as before mentioned; in others, as at Nyat and Pekin, the peat is reduced to a mass of about the consistency of mortar, spread upon the ground, cut into blocks, and left to dry and cure in the open air: in this manner a very fair article is made, and can be produced in considerable quantities, and probably at no great expense.

At Lexington, Mass., operations were commenced in 1864, under the patents of Ashcroft and Betteley, and have been continued during three seasons. Their process, as claimed, provides for separating the fibrous from the thoroughly decomposed portions of the peat by combing; in doing which the mass is reduced to a pulp, which is then conveyed into high tanks, where it is proposed to allow it to remain, until, by its own weight and pressure, it shall have become sufficiently dense to be formed into blocks, when, by opening a small gate near the bottom of the tank, it is presumed that the pressure of the superincumbent mass will force it out in a continuous sheet of uniform size, as regulated by the orifice, which may then be cut in blocks, and laid away to dry. This, as we understand it, is the theory.

They have also experimented considerably upon a method of drying the peat by absorption; the plan being to cover the spreading ground with a layer or pavement of porous brick, arranged in convenient manner, on which is spread the soft pulp as it comes from the machine. The brick, if dry, will undoubtedly absorb a portion of the moisture from the under side of the mass, while evaporation is going on from the top; but it is equally obvious that the bricks will unavoidably absorb from the earth, gravel, or sand, on which they are laid, a very considerable amount of moisture, and that this, with an occasional shower, would do much towards keeping them so well filled with water, that they could not be expected to be in fit condition for effectual service in the manner intended, except for occasional and limited periods. The expense to be incurred for paving an area sufficient to meet the requirements of a large establishment would be more than for buildings and machinery.

If an absorbent is to be used at all, it should be constantly available and comparatively inexpensive.

Operations at these works have been prosecuted vigorously, a great deal of expensive machinery and apparatus has been built, a variety of experiments have been tried, some good fuel has been produced, and a large amount of money has been expended; but we do not learn that the desired results have yet been attained.

The machinery set up at Pekin, N. Y., in 1865, is the invention of Mr. M. S. Roberts, who is the owner of a very considerable tract of excellent peat-land in that place. The following description of it is from the "Buffalo Express," of Nov. 17, 1865: "In outward form, the machine was like a small frame house on wheels, supposing the smoke-stack to be a chimney. The engine and boiler are of locomotive style; the engine being of thirteen-horse power. The principal features of the machine are a revolving elevator and a conveyer. The elevator is seventy-five feet long, and runs from the top of the machine to the ground, where the peat is dug up, placed on the elevator, carried to the top of the machine, and dropped into a revolving wheel that cuts it up, separates from it all the coarse particles, bits of sticks, stones, &c., and throws them to one side. The peat is next dropped into a box below, where water is passed in, sufficient to bring it to the consistency of mortar. By means of a slide under the control of the engineer, it is next sent to the rear of the machine, where the conveyer, one hundred feet long, takes it, and carries it to within two rods of the end; at which point the peat begins to drop through to the ground to the depth of about four or five inches. When sufficient has passed through to cover the ground to the end of the conveyer, — two rods, — the conveyer is then swung round about two feet, and the same process gone through as fast as the ground under the elevator, for the distance of two rods in length and two feet in width gets covered, the elevator being moved. At each swing of the elevator, the peat just spread is cut into blocks (soft ones, however) by knives attached to the elevator. It generally takes from three to four weeks before it is ready for use. It has to lie a week before it is touched, after the knives pass through it, when it is turned over, and allowed to lie another week. It has then to be taken up, and put in a shed, and within a week or ten days can be used, although it is better to let it remain a little longer time. The machine can spread the peat over eighteen square rods of ground — taking out one square rod of peat — without being moved. After the eighteen rods are covered, the machine is moved two rods ahead, enabling it to again spread a semicircular space of some thirty-two feet in width by eighteen rods in length. The same power which drives the engine moves the machine. It is estimated by Mr. Roberts, that, by the use of this machine, from twenty to thirty tons of peat can be turned out in a day." Four men are required to run it. The cost of this machine is stated at twenty-five hundred dollars.

In this case, as in several others which have come to our notice, it is observed that the method proposed requires that a very considerable quantity of *water* be *added* to the already moist material, before it can be treated or formed; whereas a great desideratum has ever been to discover some process by which it might be discharged of the very large amount of water which all peats contain in their natural state.

Mr. James Hodges, of Montreal, after considering the many difficulties in the way of manufacturing peat successfully, conceived the idea of a manufactory complete, which might be made to float about in the bog, excavating, pulping, manufacturing, and spreading out the pulped peat to dry, until some seventy per cent. was evaporated, or it was fit for carriage to the store or to market. After three years' experience, he has arrived at the conclusion that it may be effected in the following manner: —

"An extensive undrained bog, from eight to twelve feet in depth, — or, if deeper, the better, — having been selected, the first process is to trace out, at some distance from the margin, a contour level line of say several miles in extent. Along this line, a space of some nineteen feet in width must be cleared, and the live moss or turf entirely removed: by the side of this a space ninety feet in width is to be cleared and drained to receive the pulped peat.

"At one end of the contour line before-mentioned, a barge or scow eighty feet long, sixteen feet beam, and six feet deep, must be constructed, and launched into a hole dug in the bog to receive her. The barge or scow is to contain all the machinery necessary for the complete manufacture of the peat.

"At one end of the scow are placed a pair of large screw augers eleven feet in diameter, which, being provided with proper shafting and gearing, are made to revolve by means of a steam engine placed on the rear of the vessel. These augers or screw excavators bore out the peat in precisely the same manner that a common auger bores itself into wood; and the scow being made to move onwards as the boring proceeds, it follows that a canal nineteen feet wide, of from four to six feet deep, is formed, in which the scow, with her burden of machinery, floats, the water from the adjacent peat draining into and filling the canal as fast as it is made; the usual speed of the scow being some fifteen feet per hour.

"A competent engineer should determine and lay out the canal level, as well as arrange its water supply, upon which depends in a great measure the successful working of the whole.

"The peat, when bored out or excavated by the screws, is delivered into the barge, and conveyed by means of an elevator to a hopper, into which it is tumbled. It then passes through machinery which removes all sticks and roots, and, eventually destroying the fibre, reduces the peat to a homogeneous mass of soft pulp, like well-tempered mortar.

"This pulp then passes into a long spout or distributor, which, extending at right angles over the side of the scow, spreads out the pulp upon the levelled moss by the side of the canal, in a thin slab nine inches in thickness and ninety feet in width.

"After the slab of pulp has been deposited for a couple of days, or in hot weather for a shorter period, it begins to consolidate, and show symptoms of cracking. Immediately any cracks make their appearance, it must be marked out by drawing a framework, carrying curved knives, placed six inches apart, across it. A few days more hardens the pulp, so that by the aid of boards a man can walk on it, and mark it longitudinally with cuts eighteen inches apart.

"In about a fortnight the shrinkage of the pulp-slab causes the cuts made in it to open, and the whole presents the appearance of an immense floor covered with bricks eighteen inches long by six inches wide. As soon as the bricks are sufficiently hard to bear handling, they are separated and "footed;" that is, stood up on the ends, five in a stook, with one across the top, in which position they remain until dry enough to be removed to the store or to market.

"In the manufacture of peat-fuel considerable experience is required, and unless attention is paid to matters of detail, apparently of little importance, serious loss may be the result.

"In forming or uncovering the canal track, nothing more is required than that the turf or live moss, about six inches in thickness, together with the roots of all trees upon the surface of the bog, should be removed;

and, as upon all undrained bogs, the roots of such stunted trees as grow there are all on the surface, this operation is easily accomplished.

"In the preparation of the pulp-beds great care is required, and a surface should be obtained as level and even as possible. The roots of all trees must be removed; and this is more readily accomplished with the trees themselves, by which means considerable labor may be saved, one man pulling them down on one side, while another with an axe cuts the lateral roots at some distance from the stem, leaving the smaller portions behind. The long grass, shrubs, and rank mosses are cut down with a short scythe, and used in filling up any irregularities on the surface. Drains from nine to twelve inches deep should also be cut and covered over with the spare turf taken from the canal track. The soil from the drains may also be used in levelling and filling up inequalities in the pulp bed. In some places where the growth of shrubs has been very rank and coarse, the turf upon the whole surface of the pulp-beds has been cut into strips and inverted; but it is better to cut drains, and leave the turf in its natural position. The soft pulp, when poured upon it in a semi-fluid state, advances, lava-like, pressing down any small branches of shrubs and the long grasses which may be standing in the way of its onward progress.

"The pulp should not be deposited nearer than five feet of the canal, and upon this space may be placed any surplus moss or turf from the uncovering of the canal track, which will not only keep the pulp in place, but also form a road and towing-path for the canal. At the rear, or ninety feet from this bank, a double thickness of turf is all that is necessary to complete the pulp-beds.

"The canal track and pulp-beds being prepared, and the scow with its machinery in position, nothing more is required than to set it in motion, giving the necessary feed, say one and a half inches for each revolution of the screw excavators, which may be increased to three inches or more if necessary. As the screws revolve, they cut off continuous slices of the peat, which, by the assistance of a couple of men, are delivered through the rear of the shield the screws work in, into a well in the bow of the scow. These men also remove any large masses of extraneous material, such as pieces of wood, roots of trees, &c., which may work in. It is sometimes required, when working in peat which is very full of roots, to have a man placed in front to remove them, as they are brought up by the knives of the screws, roots as much as a man can lift being occasionally excavated.

"After the peat is delivered into the well, it is carried by means of an elevator and tumbled into a hopper, from which it passes through the stick and fibre catcher, the pulping and distributing trough, without any assistance whatever, it being only necessary to see that the stick catcher is kept clear, and occasionally, when the pulp is too stiff or dry, to turn on a pump until it is reduced to a proper consistency.

"The levelling of the pulp should be done as evenly and as smoothly as possible. A few days' experience will enable any intelligent man to accomplish this; and upon its being well done depends, in some measure, the quality of skin upon the peat, so essential, not only in shedding the rain and preventing cracking from the sun, but also for giving a permanent toughness to the bricks.

"The crew of the scow, all told, will number six, including the master, who keeps the knives of the screw excavator clean, and sees that all is going on right; two men at the screw excavators, one engine-man, one man levelling the pulp, and one man to attend to the stick-catcher and the pulping-spout.

"The marking of the pulp-beds into transverse cuts, at six-inch intervals, is proceeded with as soon as the pulp begins to set, or becomes so tough that when the incisions or cuts are made in it by the knives, they do not re-unite. The operation is performed by two men, one on each side of the pulp-bed, who, by means of a rope, pull a framework of wood, carrying curved knives, to and fro across the bed. A little practice enables them to perform the work with great accuracy. The longitudinal cuts, eighteen inches apart, are made as soon as the pulp is sufficiently hard to bear the weight of a man upon a plank laid on its surface. It is performed by pushing a circular plate of iron, which, cutting like a circular saw, severs the peat to the very bottom. In making these last cuts, care should be taken that they go quite through the peat, so that surface water from rain may freely pass off through the drains in the pulp-beds into the canal.

"Upon the state of the weather depends the time when the next operation should be performed; but, if the pulp-slab, when first spread out, is not more than nine inches in thickness, — which it should never exceed, — then a fortnight will be ample time to harden the bricks for footing.

"The footing is done by gangs of men and boys, one man and three boys working together: the man, using a suitable tool, separates the bricks, which the boys foot, or place in groups or stooks of five; four stand on their ends, inclining to each other, with their tops touching, the fifth being balanced horizontally upon them. A man and three boys will foot four thousand bricks in a day.

"After the bricks have been exposed to the weather for a few days, they should be refooted or turned, two boys handling four thousand as a day's work.

"Nothing now remains to be done but to wheel the bricks, when sufficiently dry, into barges, and convey them to the store."

Mr. Hodges' plan of operations is rather extensive, not to say immense; but he has had — what at this stage of affairs we should rarely expect to see — the requisite enterprise, energy, and perhaps the location and pecuniary means to carry it out, and has made clear and convincing statements of results attained, which are of great importance as demonstrating the comparative value of this fuel. We have been favored with reports of some of these, which will be found on subsequent pages, under the heads of "Steam" and "Iron."

Several trips with his fuel were made over the Grand Trunk Railway, which were all attended with marked and satisfactory results, so decided in their character that a contract has been entered into, extending over five years, or seasons, during the first of which the company are to take one hundred tons per day, and during the four succeeding seasons three hundred tons per day.

Dr. Louis Elsberg, of New York city, obtained a patent, in 1864, for a process of grinding and compacting dry peat, which in its essential features appears to correspond very nearly with the Lithuanian, Hodgson's, Gwynne's, and Exter's, already described on pages 44, 45, 46; and the remarks in connection with those would seem to be equally applicable to this, so far as we are able to judge from our own observations and the concurrent testimony of others. His experimental works are at Belleville, N. J., and the fuel produced is in good shape and very dense.

The Boston Peat Company have adopted the process and machinery invented by T. H. Leavitt, of Boston; and they are in successful operation.

The process is exceedingly simple, rapid, and successful: the machinery is equally simple, and works "to a charm." It is compact, and of moderate cost; and the results obtained in the quantity, quality, and cost of the fuel produced, exceed the most sanguine expectations of the inventor, and have met the unqualified approval and commendation of all who have examined the machinery or the fuel.

The machinery consists of a strong tank, or cistern, three feet in diameter and six feet high, supported upon a stout framework, about four feet above the floor of a suitable building, which should be near the bog, and is best constructed on a hill-side, so that easy access can be had to the lower story on one side, from the base of the hill, and to the second story on the other side. The top of this tank is open, and even with the floor of the second story. Within the tank, and firmly fixed to its sides, are numerous projections of a variety of forms, adapted to the treatment of the material in its several stages as it progresses through the mill, which is divided into three apartments: through the centre of the tank revolves an upright shaft, to which are affixed knives and arms varying in form and structure to correspond with the stationary projections in each apartment; below the tank is a receiver, or hopper; and under this is a moulding or forming machine, two feet in width and twelve feet long, of like simple construction, which receives the condensed material from the hopper, and delivers it in blocks of any desired form and size. The whole is adapted to be driven by a small steam engine, and requires about six and ten horse power, respectively, for the two sizes of machines, as at present constructed, of the capacity of fifty and one hundred tons each of crude peat per day of ten hours.

The crude material is brought from the bog in ordinary horse-carts, or on small cars running over a cheaply constructed tramway, to the mouth of the mill, in the floor of the second story of the building, where it is dumped or shovelled into the mill in any convenient quantity; but the arrangement is such that only a given amount is admitted and under treatment at any one time, so that all parts have a uniform and regular supply. The treatment is such that the original organization of the peat is entirely destroyed; in the second stage, the air, of which a large amount is contained in its cells, is ejected: advantage is taken of some of the natural properties of the material, and the mass is condensed in its moist state in the lower part of the mill, from whence it is delivered into the hopper of the moulding machine, and is discharged in a continuous line of moulds (which are fed into the rear part of the machine by a boy), at the rate of from fifty to one hundred tons per day of ten hours. The work of removing the blocks to the spreading-ground is easily accomplished; and they are exposed in the open air, for drying, in much the same manner as bricks are exposed in a brick-yard.

The amount of water contained in well-drained peat is ordinarily from 65 to 75 per cent., varying according to the character of the material and the drainage of the meadow; so that the weight of dry, hard fuel from the product of a day's operations, is from 12 to 17 tons, or 25 to 35 tons, from the two sizes of machines, respectively, —the cost of which, by this process, to place it on the spreading-ground, at present prices of labor ($1.75 per day), is less than $1.50 per ton; to which may be added 50 cents per ton for turning the fuel while drying, and for housing it; making the entire cost $2 *or less.*

The water remaining in the blocks as they come from the mill can be got rid of, only by evaporation, which goes on very rapidly after this method of treatment; and the fuel is, at the expiration of about eight or ten days, — sometimes in four or five, — in condition to be housed, or transported to market.

The cost, as stated above, is for the product of one set of machinery; but where several machines are to be operated, and the business is conducted on an extensive scale, the cost, *pro rata,* is very much reduced, as one man can easily superintend the operation of several machines, the laborers generally can be employed to better advantage, and numerous mechanical appliances to save manual labor and expedite the operations, which it would not be advisable to construct where a single machine only was to be run, may be economically introduced on more extensive works; and in this manner the expense may be very considerably reduced, some have estimated as low as one dollar per ton.

The parties interested in this latter enterprise have been diligently pursuing the matter for a long time, firmly convinced of the value of the article, and of the practicability of producing it, in marketable shape, in large quantities, and at moderate cost; and are content that the merits of their machinery, and the value of the product as an article of fuel, should stand upon the test which each observer or consumer may choose to apply.

Their inventions are secured by letters-patent, and the process can be seen in operation at their works, at East Lexington, Mass.

Their purpose is to encourage the manufacture of peat wherever it is found; and to this end they furnish machinery, and rights under their patents, at moderate rates.

The following is a liberal estimate of the cost of an establishment for running a Leavitt machine of the capacity of one hundred tons crude peat per day, estimating labor and material to command about the prices which rule at the present time, 1867, which are high: —

One machine, as above,	$1,500
Engine and boiler, 12 horse,	1,200
Shafting, belting, and fixtures,	500
Buildings, roughly constructed,	1,000
Incidentals,	300
	$4,500

Or for a better establishment, as follows: —

One machine, as above,	$1,500
Engine and boiler, 12 horse,	2,000
Shafting, belting, and fixtures,	500
Buildings,	2,000
Incidentals,	500
	$6,500

For machines of fifty tons capacity, the cost is $500 less on each machine; the power required is, of course, less, and the cost for an engine would be reduced proportionately.

The number of machines to be run in one establishment may be increased with comparatively small outlay for buildings and power.

The laborers employed, and the cost of labor for running one of these machines, will doubtless be an item of interest to those investigating the subject. At the works of the Boston Peat Company, at Lexington, Mass., they have been as follows, for 100 tons crude peat per day:

Four men to cut the peat from the bog and load it into cars; one boy, one horse, and two cars, to haul (on a tramway) from the bog to the mill; one man to feed into the mill; one man to put moulds into the mill; two men to take moulds from the mill; one boy, one horse, and three trucks to remove the moulds (filled with peat) to the spreading ground; two men on the spreading ground to empty the moulds; one boy to receive the moulds as they come back from the spreading ground; one engineer and one superintendent.

This comprises 12 men, 3 boys, and 2 horses.

Ten of the men are paid .	$1.75	per day . .	$17.50
The engineer is paid . .	2.50	" " . . .	2.50
The superintendent is paid	3.00	" " . . .	3.00
The three boys are paid .	75	" " . . .	2.25
For the two horses we pay	1.25	" " . . .	2.50
			$27.75
In addition to this the blocks of peat require to be "haked" or turned once while on the spreading ground, which is done by boys, and the dry fuel is to be gathered up and placed under shelter, which can easily be done by two men, and occasionally an extra man or boy may be wanted about the premises; all of which is considered to be more than covered by ten dollars per day, ...			10.00
Giving as a total cost for labor on each day's yield of, say 25 tons or more, of merchantable fuel,			$37.75

In this connection, some data concerning peat-beds,

and the product of manufacture, may be of interest. We therefore give the following as some of the results of our own observation, from repeated practical tests at the works of the Boston Peat Company, located at East Lexington, Mass., about ten miles from Boston.

A cubic foot of crude peat, as taken from a well-drained bog, weighs from fifty to fifty-five pounds.

This same quantity is *condensed* by the machinery in use at the works above mentioned, from thirty to forty per cent., according to the character or structure of the material, and that too, before it is relieved from any of the water contained in the mass.

In this state it is formed into blocks of convenient size, which are then exposed in the open air, where evaporation takes place very rapidly, and is found to be greatly accelerated by the treatment the material has received in passing through the condensing mill; so that the time ordinarily required in the summer season for drying sufficiently to be housed is about six or eight days, though it varies from four to ten days, according to the weather; and by evaporation it is still further reduced to about one quarter its original bulk, and will, at this stage, have been found to have lost about two thirds to three quarters of its original weight, its *bulk* having been diminished by the forcible ejection of the air in the process of condensing, and the loss of water by evaporation, while its *weight* is diminished solely by the evaporation of the water.

A ton of wet peat, as cut, will measure about forty cubic feet; and about one hundred and sixty cubic feet of crude material are required to produce one ton of dry fuel. Some very compact peats, however, require not more than 140, or even 120, cubic feet for a ton of dry fuel.

One block, as it comes moist from the moulds, measures 8 × 4 × 2½ (80 cubic inches), and weighs about 3 lb. 6 oz.

The machines turn out respectively 30,000 and 60,000 of these blocks per day.

The best place to dry the blocks is on the grass, in the open air, where they dry most rapidly and uniformly.

A very economical method of drying is, undoubtedly, on racks, consisting of a light frame about seven feet long by three feet wide, crossed with ordinary laths, convenient to receive 48 blocks each (8 × 6=48), and are easily handled by two men. These may be piled one upon another, while good ventilation is secured, and the cost of handling is less than by any other method, except upon the grass, as above.

Of those blocks, wet as they come from the mill, 610 weigh one ton.

As these blocks are laid out to dry, an acre of land will be covered by about 250 tons — spread on the grass.

If spread on racks or frames, as above described, one ton will require 13 frames.

One acre of land, then, allowing one fourth the area for drive-ways, &c., will accommodate 1650 frames (one deep), or about 125 tons.

The same area, covered with frames 10 tiers high, will accommodate 1250 tons.

Peat can be manufactured and dried in this manner, in good weather, so as to make a good fuel, and be suitable to burn, in a week or ten days; but, like good dry wood, it is undoubtedly much improved in quality when housed and allowed to cure for a season, say three to six months.

An acre of peat, of fair quality, well drained, if one

foot in depth, will generally contain 1000 to 1200 tons, yielding 250 to 350 tons of dry fuel. Greater depths, in proportion.

Few peats, however well drained, contain less than 50 per cent. of water, and most contain 65 to 85 per cent.

Our own estimates have always been made on 75 per cent. of moisture, which is safe; but it is quite probable that 70 per cent. would be fair in the majority of well-drained meadows.

When best drained, peat is worked to the best advantage, both as regards economy of labor and the quantity and quality of fuel produced.

Peats vary much in their heating properties, as do woods and coals, according to their growth and composition. The most thoroughly decomposed and compact deposits yield, when manufactured, the most dense fuel at no greater cost for labor than the lighter and more porous material, and are, therefore, the cheapest.

It is an essential feature of this process, and one which, in some sections, will be found of great importance, that from light and inferior qualities of crude peat, we are able to produce an excellent article of fuel.

Pure moss peats are invariably good. The most resinous peats are shown to be the most valuable, especially for generating steam and for the production of gas.

As to the specific gravity of condensed peat: we have often heard it said that "peat is equal to the best hard wood." Now, we know that the best and hardest woods will float upon water; while it is a fact that, from the very poorest peats we have ever worked, our machinery has never failed to produce a fuel which would sink in water; showing its specific gravity to be greater than the best of woods.

We are able to produce fuel varying from 65 to 80 pounds per cubic foot, according to the character of the crude material, which is equal to from 4 to 5 tons to the solid cord, or $2\frac{1}{2}$ to $3\frac{1}{2}$ tons in its broken condition, as shovelled up and loaded when dry.

In estimating the weight of dry peat-fuel per cubic foot, bushel, or otherwise, it should be understood that the *waste* space in peat and in wood, as commonly heaped or piled, is probably not far from 40 per cent.

The quality of peat-fuel, like wood, is improved by age; that is, the fuel, although housed very soon after it is manufactured, and considered dry, and in excellent condition for use, as it really is, will be found to have improved very much if properly housed, and allowed to remain and cure for three, six, or even twelve months, the difference in quality being as readily observed as in wood, under the same circumstances.

Although peat-fuel, properly manufactured, will stand considerable exposure to the weather, it will inevitably be injured, to some extent, by rain and snow, sun and frost, if left uncovered long after it is fit for use; and every manufacturer and consumer will find it to be the wiser course, if he has a good article of fuel, to provide a suitable place for it, and take good care of it.

Frequent inquiry is made as to the practicability of drying peat by artificial means, and the best method of accomplishing it. That it *can* be dried by artificial means has been satisfactorily demonstrated; but we are by no means prepared to say that the *best* and most economical method for accomplishing it has yet been devised. A great variety of kilns are in operation, some of them certified and acknowledged to work almost to perfection in drying lumber, cotton, tobacco, &c.; but wet peat is obstinate, will not yield kindly to the same treatment, in fact is "a poser." Its peculiarities have yet to be fully understood.

The points to be considered in perfecting a process for artificial drying, aside from the first cost of buildings and apparatus, are, the time required, the expense for fuel and labor, and the quality and characteristics of the fuel produced, as affected by the manner of drying.

We have no idea that artificial drying will be resorted to, to any extent, during good weather in the summer, at present, for Nature accomplishes the work for us, when she does it at all, better than Art can, and at much less expense; but in stormy weather, and during the winter season, we are satisfied that it can be done; and although the best method has not yet been proved, we are on the direct road to it, and, by patient, persevering effort, with a willingness to "make haste slowly," we fully expect to reach it in good time.

A light current of heated air, passing over and through the mass of peat, is what is required; the details of buildings, mechanical arrangement of the apparatus, and cost of fuel and labor, by which to attain the best results in the most rapid and economical manner, are points which cannot be said to have been sufficiently elucidated to admit of writing definitely in regard to them.

Steady progress is being made, and time, skill, and enterprise are sure to demonstrate the *best* method of artificial drying.

PEAT-CHARCOAL.

Not merely may we utilize peat in its natural condition, or in its manufactured and solidified state, but we may carbonize it as we do wood, and produce peat-charcoal.

The common and simple mode of carbonizing or charring the ordinary peat is in heaps, in the same manner as that of wood. The sods or blocks must be regularly arranged, and laid as close as possible: they are the better for being large, — say fifteen inches long by six broad and deep. The heaps, built hemispherically, should be smaller in size than the heaps of wood usually are. In general, some five thousand or six thousand large sods may go to the heap, which will thus contain about fifteen hundred cubic feet. The mass must be allowed to heap more than necessary for wood; and the process requires to be very carefully attended to, from the extreme combustibility of the charcoal. The quantity of charcoal obtained by this method is generally from twenty to thirty per cent. of the weight of dry turf.

For many industrial uses, however, the charcoal so produced from peat in its natural state is too light; because, generally speaking, it is only with fuel of considerable density that the most intense heat can be produced.

It is, therefore, only from peat in a manufactured or solidified state that we can expect to prepare a charcoal thoroughly adapted as a fuel for the more severe processes required in the arts.

By coking this, however, a charcoal is produced of a density of 1.040 or upwards, which is far superior to the best wood-charcoal, and is fully equal to that of the best coke made from coal.

Its calorific power is *intense*. The quality of charcoal generally obtained from a good article of solidified peat is from thirty-five to forty-two per cent. of its weight.

In the great Exhibition at Paris, in 1851, numerous specimens of peat and peat-charcoal, prepared by different patented processes, were exhibited, which were remarkable for their density and cheapness, and attracted particular attention. They were stated to be economically employed for stationary steam engines and for locomotives.

Moulded peat, in small bricks, of density sufficient to sink in water, is supplied in the city of Paris from numerous sources, mostly for domestic purposes. From Liancourt, distant seventeen leagues, it is brought, and sold at the rate of twenty francs for 2204 pounds avoirdupois. It is stated of one firm, that, in 1855, they converted some ten thousand tons or more into charcoal, obtaining from forty to forty-two per cent., which was sold at wholesale for one hundred francs the one thousand kilograms (two thousand two hundred and four pounds), which was about the same value as wood-charcoal, and about three times the price of wood and mineral coal for the same weight.

The methods adopted for charring or carbonizing peat differ somewhat in the various localities where it is prepared. We have seen descriptions of the manner in which it is conducted in England, Ireland, France, Bohemia, Bavaria, Saxony, Russia, Friesland, and elsewhere; the general principles being the same, but differing in details, some of which appear to offer decided advantages in an economical point of view, and in the production of the greatest amount of compact fuel from a given weight of the raw material.

From Prof. Johnson we quote: —

"When peat is charred, it yields a coal or coke which, being richer in carbon, is capable of giving an intenser heat than peat itself, in the same way that charcoal emits an intenser heat in its combustion than the wood from which it is made.

"Peat coal has been and is employed, to some extent, in metallurgical processes, as a substitute for charcoal, and, when properly prepared from good peat, is in no way inferior to the latter — is, in fact, better.

"It is only, however, from peat which naturally dries to a hard and dense consistency, or which has been solidified on the principles of Challeton's and Weber's methods (condensed in its *moist* state), that a coal can be made possessing the firmness necessary for furnace use. Fibrous peat, or that condensed by pressure, as in Exter's, Elsberg's, and the Lithuanian process, yields, by coking or charring, a friable coal, comparatively unsuited for heating purposes.

"A peat which is dense as the result of proper mechanical treatment and slow drying, yields a very homogeneous and compact coal, superior to any wood charcoal, the best qualities weighing nearly twice as much per bushel.

"Peat is either charred in pits and heaps, or in kilns. From the regularity of the rectangular blocks into which peat is usually formed, it may be charred more easily in pits than wood, since the blocks admit of closer packing in the heap, and because the peat-coal is less inflammable than wood-coal. The heaps may likewise be made much smaller than is needful in case of wood, viz., six to eight feet in diameter, and four feet high.

"I have carbonized, in an iron retort, specimens of peat prepared by Elsberg's, Leavitt's, and Ashcroft and Betteley's processes. Elsberg's gave 35, the others 37 per cent. of coal. The coal from Elsberg's peat was greatly fissured, and could be crushed in the fingers to small fragments. That from the other peats was more firm, and required considerable exertion to break it. All had a decided metallic brilliancy of surface."

The high heating power of peat-charcoal, and its freedom from properties deleterious to metal, must invest it with peculiar interest to the smelter, and those who follow after him as manufacturers of the article he produces; while, to the sanitary reformer and the agriculturist, its disinfecting and fertilizing qualities may be said to be of hardly less importance.

PEAT IN EUROPE.

The abundance and accessibility of peat in Ireland render it of no small importance among the natural resources of that land, especially to the vast mass of the poorer classes. It is generally of superior quality, plentiful, and cheap. Not only is it the common fuel of the poor in the interior, — and, indeed, of all classes, in some districts, — but it is transported in barges, in immense quantities, by canal, to Dublin, and there consumed by the wealthier class of the people.

So extensive is the supply of peat in Ireland, that it has been estimated to occupy one seventh of its entire surface. It is stated that Ireland has two canals running through two million acres of peat-bog. Among other instances of the value of peat as an article of fuel, it is stated that a distillery company, by the judicious management of a bog, had their steam power for half the cost required for coals, and were, at the same time, making an estate of reclaimed land for themselves.

The red peat-bogs, which form so remarkable a feature in this country, are chiefly comprised in the great central plain of Ireland. Unlike the English mosses, they are rarely level, but undulating; and, in Donegal, there is a bog which is completely diversified by hill and dale.

These bogs consist of moist vegetable matter, containing a great deal of stagnant water, and, after heavy rains, have been known to burst, and inundate the adjoining country.

At a meeting of the British Association, in 1842, Mr. Griffith illustrated the mode in which he considered the coal-measures had been formed, by describing the general condition of the peat-bogs of Ireland. They appeared to occupy basins which had formerly been lakes; but the peat-moss had grown up to the level of the water, and afterwards, by capillarity, had risen twenty or thirty feet higher. As a case in point, he mentioned a bog, the base of which consisted of clay, covered by a layer of peat which is composed of rushes and flags. Above this is another bed of peat, closely resembling cannel-coal, "possessing a *conchoidal fracture, and hard enough to be worked into snuff-boxes*." It yielded twenty-five per cent. of ashes, and contained a large proportion of oxide of iron.

This bed was covered with black peat, containing branches and twigs of fir, or pine, oak, yew, and hazel, only the bark being left; and, where whole trees occurred, the roots were entirely gone. The surface was formed of ordinary bog-moss (*sphagnum*), nearly white. The whole amount of peat in the bog referred to, would, he thought, form a coal-seam of at least three or four feet in thickness.

We have seen several statements, of late years, to the effect that the area of peat-land in Ireland is considerably diminished; some of the bogs having been reclaimed, and converted into arable land, and others exhausted, drained, or cut out.

The Bog of Dourah, eastward from the Fergus, affords the principal supply of peat to Ennis and Clare. The bogs in this district abound in timber. A fir-tree, measuring thirty-one to thirty-eight inches in diameter, by sixty-eight feet in length, is mentioned as having been raised from a bog near Kilrush. The mode of finding bog-timber is rather remarkable. It is ascertained that the dew does not lie on that portion of a bog immediately above a tree or log, as it does elsewhere. Its position can thus be easily ascertained before the dews rise in the morning; when the finder, after probing with a bog-auger to ascertain whether the wood be sound, marks the spot with a spade, and proceeds to raise the timber at his leisure. Much of this timber is sound; and from some bogs very large and exceedingly valuable sticks have been raised, the growth of past ages.

The series of extensive bogs in the central part of Ireland, though separated from each other, have received the common name of "The Bog of Allen." They vary much in depth, composition, moisture, &c. They rest, generally, upon a stratum of blue clay, based on limestone, and are invariably above the level of the sea. Their greatest elevation, however, does not exceed four hundred and ninety feet; the mean elevation being two hundred and fifty feet.

The Parliamentary Commission, appointed about 1812, to inquire into the nature and extent of the several bogs in Ireland, and the practicability of draining and cultivating them, reported, in 1814, that "the extent of peat-soil in Ireland exceeds two million eight hundred and thirty-one thousand English acres," of which there were at least one million five hundred and seventy-six thousand acres of flat red-bog, considered the most valuable; the remainder consisting of mountain-bogs, on the surface of the uplands.

The drainage and cultivation of these extensive proportions of the surface of Ireland have long been regarded as objects of great national importance. Numerous commissions have been appointed to investigate and report, and frequent attempts have been made to show that these ends might be attained at no very great expense; but the instances of successful bog cultivation in any part of the island are very few, and those only under peculiar circumstances. The bogs have a value which none can question, even though they be entirely useless for purposes of cultivation. They supply the inhabitants extensively with fuel. In those parts, indeed, where bogs are scarce, they are the most valuable properties in the country. In some localities, the peat has been entirely cut out; and where this is the case, and other bogs are not easily accessible, the inhabitants have sustained great privations from the want of fuel.

It has been remarked, that "the rainy climate of Ireland, and the wet occupations of the people, together with the nature of their food, render a fire more essential to their welfare than to most others; and, in fact, it is frequently the substitute for clothing, bedding, and, in part for shelter. Had it not been for the bog, the measures taken in former times to extirpate the nation might probably have succeeded; but the bog gave them a degree of comfort upon easy terms, and enabled them to live under severe privations of another kind."

We have seen an estimate somewhat as follows, intended to show how important to Ireland are her peat-bogs in furnishing a valuable fuel, independent of her deposits of anthracite and bituminous coal.

The quantity capable of being cut for fuel may be taken as low as two million acres, at an average depth of three yards; the mass of fuel which they contain, estimated at five hundred and fifty pounds per cubic yard when dry, amounts to the enormous quantity of 6,338,666,666 tons.

Taking, therefore, the value of peat (crude) as compared with that of coal (said to be as one to six), the total amount of peat fuel in Ireland is equivalent to four hundred and seventy million tons of coal; which, at twelve shillings per ton, is worth about £280,000,000 sterling, or $1,335,000,000.

In regard to the trees which are so frequently found in the Irish bogs, Mr. Aher remarks, "Such trees have generally six or seven feet of compact peat under their roots, which are found standing as they grew; evidently proving the formation of peat to have been previous to the growth of the trees." In the bogs in the vicinity of Londonderry, according to the Report of the Ordinance Survey in 1837, the fact above stated may be verified in relation to *fir-trees*, the lowest layer of which is underlaid by from three to five feet of peat. Not so, however, with *oaks*, as their stumps are commonly found resting on the gravel at the base, or on the sides of the small hillocks of gravel and sand which so often stud the surfaces of bogs, and have been aptly called "islands" by Mr. Aher, and "hummocks" by other writers. It is a remarkable fact, although very common, that successive layers of trees or stumps in the erect position in which they had grown, and furnished with all their roots, are found at distinctly different levels, and at small vertical distances from each other.

The bogs contain, it has been ascertained, two important families of trees, the resinous or coniferous trees, which grew in successive layers or tiers upon the ancient surfaces of peat; and the hard-wooded, non-resinous trees, which grew upon the gravel at the original base. Of the former, the prevailing tree was the common

Scotch pine, or fir, — *Pinus sylvestris:* of the latter, the oak, *Quercus robur*, prevailed.

It may be mentioned here, as a matter of some interest, that in a "notice of a submarine forest in Cardigan Bay, North Wales," the author remarks on the occurrence therein of the *Pinus sylvestris*, although the Scotch fir is now excluded from the native flora.

Professor Lyell, in his "Principles of Geology," says, —

"It is a curious and well-ascertained fact, that many of the mosses (bogs) of the north of Europe occupy the place of immense forests of pine and oak, which have, many of them, disappeared within the historical era. Such changes are brought about by the fall of trees, and the stagnation of water caused by their trunks and branches obstructing the free drainage of the atmospheric waters, and giving rise to a marsh. In a warm climate, such decayed timber would immediately be removed by insects or by putrefaction; but, in the cold temperature now prevailing in our latitudes, many examples are recorded of marshes originating in this source. Thus, in Mar Forest, in Aberdeenshire, large trunks of Scotch fir, which had fallen from age and decay, were soon immured in peat formed partly out of their perishing leaves and branches, and in part from the growth of other plants. We also learn that the overthrow of a forest by a storm, about the middle of the seventeenth century, gave rise to a peat-moss, near Lochbroom, in Ross-shire, where, in less than half a century after the fall of the trees, the inhabitants dug peat. Dr. Walker mentions a similar change, when, in the year 1756, the whole Wood of Drumlaurig was overset by the wind. Such events explain the occurrence, both in Britain and on the Continent, of mosses where the trees are all broken within two or three feet of the original surface, and where their trunks all lie in the same direction.

"Nothing is more common than the occurrence of buried trees at the bottom of the Irish peat-mosses, as also in most of those of England, France, and Holland; and they have been so often observed with parts of their trunks standing erect, and with their roots fixed to the subsoil, that no doubt can be entertained of their having generally grown on the spot. They consist, for the most part, of the fir, the oak, and the birch. Where the subsoil is clay, the remains of oak are the most abundant; where sand is the substratum, fir prevails.

"In the Marsh of Curragh, in the Isle of Man, vast trees are discovered standing firm on their roots, though at the depth of eighteen or twenty feet below the surface. The leaves and fruit of each species are frequently found immersed along with the parent trees; as, for example, the leaves and acorns of the oak, the cones and leaves of the fir, and the nuts of the hazel.

"The durability of pine-wood, which in the Scotch peat-mosses exceeds that of the birch and oak, is due to the great quantity of turpentine which it contains, and which is so abundant that the fir-wood from bogs is used by the country people, in parts of Scotland, in the place of candles. Such resinous plants, observes Dr. Macculloch, as fir, would produce a fatter coal than oak, because the resin itself is converted into bitumen.

"In Hatfield moss, which appears clearly to have been a forest eighteen hundred years ago, fir-trees have been found ninety feet long, and sold for masts and keels of ships: oaks have also been discovered there above one hundred feet long. The dimensions of an oak from this moss are given in the Philosophical

Transactions, No. 275, which must have been larger than any tree now existing in the British dominions.

"In the same moss of Hatfield, as well as in that of Kincardine and several others, Roman roads have been found, covered to the depth of eight feet by peat. All the coins, axes, arms, and other utensils found in British and French mosses, are also Roman; so that a considerable portion of the European peat-bogs are evidently not more ancient than the age of Julius Cæsar: nor can any vestiges of the ancient forests described by that general, along the line of the great Roman way in Britain, be discovered, except in the ruined trunks of trees in peat.

"De Luc ascertained that the very site of the aboriginal forests of Hircinia, Semana, Ardennes, and several others, are now occupied by mosses and fens; and a great part of these changes have, with much probability, been attributed to the strict orders given by Severus and other emperors to destroy all the wood in the conquered provinces. Several of the British forests, however, which are now mosses, were cut at different periods by order of the English Parliament, because they harbored wolves or outlaws. Thus the Welsh woods were cut and burnt in the reign of Edward I., as were many of those in Ireland by Henry II., to prevent the natives from harboring in them and harassing his troops.

"It is curious to reflect that considerable tracts have, by these accidents, been permanently sterilized; and that, during a period when civilization has been making great progress, large areas in Europe have, by human agency, been rendered less capable of administering to the wants of man. Dr. Rennie observes with truth, that in those regions alone which the Roman eagle never reached, in the remote circles of the German Empire, in Poland and Prussia, and still more in Norway, Sweden, and the vast empire of Russia, can we see what Europe was before it yielded to the power of Rome. Desolation now reigns where stately forests of pine and oak once flourished, such as might now have supplied all the navies of Europe with timber.

"At the bottom of peat-mosses is sometimes found a cake, or 'pan' as it is termed, of oxide of iron; and the frequency of bog iron-ore is familiar to the mineralogist. The oak, which is so often found dyed black in peat, owes its color to the same metal. From what source the iron is derived is by no means obvious, since we cannot in all cases suppose that it has been precipitated from the waters of mineral springs. According to Fourcroy, there is iron in all compact wood; and it is the cause of one twelfth part of the weight of oak. The heaths (*Ericæ*) which flourish in a sandy, ferruginous soil, are said to contain more iron than any other vegetable.

"It has been suggested that iron, being soluble in acids, may be diffused through the whole mass of vegetables when they decay in a bog, and may, by its superior specific gravity, sink to the bottom, and be there precipitated, so as to form bog iron-ore; or, where there is a subsoil of sand or gravel, it may cement there into iron-stone or ferruginous conglomerate."

FRANCE.

There are several deposits of peat which furnish the supply of this material for the Paris market. A portion of a large peat-bog near Liancourt, on the Northern Railway, seventeen leagues from Paris, is wrought by

Messrs. Debonne & Co., who employ about three hundred men during five months of the year. The peat here has an average thickness of about ten feet. The cuttings from the top and bottom of the bog are mixed, and, being transferred to flat-boats, is turned over with shovels and trampled by the feet of men; after which it is moulded with pressure into the form of small bricks, which, when dried, are ready for market.

The quantity annually raised by this concern is from ten thousand to twelve thousand tons, a large portion of which is converted, upon the spot, into charcoal, of which the yield is from forty to forty-two per cent.

The moulded peat is worth, in Paris, about twenty francs per ton; the charcoal, about one hundred francs per ton. This peat yields about ten per cent. of ash, and the charcoal twenty-seven per cent.; which indicates a quality inferior to most American peats.

M. Herbert, of Reims, prepares a large quantity of compressed peat of excellent quality, amounting to about fourteen thousand tons annually, a part of which is manufactured into charcoal.

The peats and charcoals prepared by the patented process of Challeton at Clermont Ferrand and Mantauger, specimens of which were shown at the great Exhibition of 1851, were remarkable for their density and cheapness, and attracted particular attention. They were said to be economically employed for stationary steam engines and locomotives.

Dr. Elisha North, of New London, Conn., writes in 1825, "I believe, judging from much experience, that peat of the best quality, if used for producing a genial and pleasant temperature in common winter weather, *is the best fuel which the earth produces*, unless an exception be made in favor of a very few species of trees equally well prepared.

"My opinion respecting its utility has been confirmed by judicious and candid persons who have been practically and thoroughly acquainted with it."

He adds, "As evidence that public opinion in some places is in favor of peat, it may be mentioned that Citizen Ribaucourt published by order of the French Government, in the eleventh year of the republic, a regular treatise on the subject. In this treatise, he says ten thousand persons are annually employed in preparing and transporting peat from one peat basin or marsh upon the lesser branch of the Loire, in the north-west part of France. This peat-bed cannot be a great distance from the city Nantes, where much peat is burnt; or even from the city of Paris."

A translation of a part, and a synopsis of the remainder, of the report of Ribaucourt, referred to by Dr. North, was subsequently published in Silliman's "Journal of Science" in 1828. From this it appears that peat and its various uses had long been known in France. Its use, however, had been limited to the departments of Somme, Loire-inférieure, Pas de Calais, some cantons of the departments of Oise, Marne, Eure, Seine et Oise, Meurthe, Vosges, and to a few others. There is scarcely a valley which does not contain valuable deposits of it, the thickness varying from six inches to twenty feet.

He remarks that peat and the coal of peat may be put to the same uses as wood and charcoal, and may be advantageously employed in a great number of the arts. It is employed not only for domestic purposes, but in furnaces under boilers, in burning brick and lime, and in preparing plaster. The ashes are very valuable in agriculture, and command a high price. Observations show that the water which has penetrated peat beds has antiseptic properties.

We find in the work of De Luc, entitled "A Letter upon Men and Mountains," some interesting details upon the progressive accumulation of peat, analogous to that of glaciers in certain mountains. There have been found under certain beds, in the Valley of the Somme, ancient causeways, divers tools, and pieces of money.

From a paper entitled "Peat as an Article of Fuel," published in Boston in 1864, we extract the following: —

"It is estimated, that, in France, there exists the enormous quantity of six thousand million tons of peat, purified and *dried* in the crude state; or, reduced to charcoal, two thousand six hundred and ten millions of tons; the heating power of which equals that of wood-charcoal.

"From the figures of the most skilful mining engineers in the empire, we find that —

				Degrees of Heat.
1	kilogram of	wood-charcoal . . .	yields	7,000
1	"	purified peat-charcoal	"	7,000
1	"	coal-coke	"	7,000
1	"	raw coal	"	5,000
1	"	raw wood . . .	"	2,600
1	"	raw purified peat . .	"	4,300

while *condensed* peat, deprived of the excess of oxygen, possesses nearly double the heating power of coal.

"Again: it is proved that the total general annual consumption in France of all kinds of mineral and vegetable fuels was as follows: —

Wood-charcoal for iron-works	667,902	tons.
" " " other purposes	472,630	"
Raw wood for iron-works	8,405	"
" " " other purposes	1,989,710	"
Coal-coke for iron-works	767,622	"
" " " other purposes	2,462,400	"
Raw coal for iron-works	1,108,252	"
" " " navigation, railroads, &c. . .	3,725,200	"
Raw and carbonized peat, actual consumption	359,319	"
Total	11,561,440	tons.

"If peat had been used in the place of these different kinds of fuel, it would have required 15,656,687 tons, raw and purified, to produce the same effect; and, at that rate, the supply of peat in France would have sufficed the empire for nine hundred years, without importing a pound of coal, and leaving her free to export annually the seven million five hundred thousand tons of coal she raised in 1862, and free likewise from the necessity of importing eight million tons of coal, as she did from England and Belgium in the same year.

"It is not believed that France is any better off in respect to her deposits of peat than the New England States; and it is safe to say, that, by a proper development of their resources in this respect, these States could soon make themselves independent of the world, so far at least as the supply of fuel is concerned, retaining within themselves all the vast sums they annually expend for fuel, besides giving employment to and enriching their own population.

"But can peat be used at less or even the same expense as other kinds of fuel for manufacturing purposes? Take the article of pig-iron. By the French engineers, it is found that the cost of working pig-iron, —

1 ton, with wood-charcoal, was	£4	11s.	0d.
1 " " coal-coke, was	2	16s.	0d.
1 " " raw coal, was	2	15s.	4d.
1 " " purified peat-charcoal, was	2	4s.	10d.
1 " " crude peat (condensed), was	1	10s.	0d."

"This is enough to prove the economy of peat for *all* purposes. The general result is thus stated: 'For domestic consumption, the economy — all conditions of heat being equal — would not be less than thirty per cent. on the cost of fires with wood-charcoal and wood, coke and raw coal; and that for large manufactories, which, on account of the quantity which they annually consume, should themselves produce this fuel, that calculation of economy would, in certain cases, be raised to sixty per cent.' "

Similar results follow the inquiry into the economy of using peat in locomotive engines. Of the eleven hundred and ninety-nine locomotive engines employed by four great companies of France, it was estimated that the whole ran nineteen million five hundred thousand miles per year. They used coal and coke concurrently, amounting in the aggregate to a hundred and eighty-one thousand tons, at a cost of £446,711 sterling. Purified peat would have done the same work for £271,500, making a saving of £175,271, to say nothing of the less amount of wear and destruction of the boilers, grates, &c.

But as to bulk, does not peat require more room? "It has been proved and acknowledged, that, for equal bulk, raw purified peat contains one third more heat than coke, and less by one fifth only for an equal bulk of coal of good quality."

Says M. Bute, *Superintendent of Railway Engineers for the Kingdom of Hanover*, "We can, by the help of a hopper placed on the tender, carry the quantity of peat which would be necessary for a trıp of one hundred to one hundred and twelve English miles. No difficulty will be presented to the employment of compressed peat for heating ordinary fixed engines, and, eventually, for that of steamboats.

"On this subject it is sufficient to say, that, by the process employed by the 'General Association for Working the Peat and Metalliferous Deposits of France,' a density may be given to peat, when condensed, of from six to fifteen hundred weight per cubic yard."

These facts and conclusions have been principally obtained from reports of companies in England, and of scientific associations in France, who have studied the subject of preparing peat, both as a money making enterprise and as an element in political economy.

Italy.

From the reports of the great Exhibition of 1862, we quote as follows: —

"Looking at the difficulties caused by the neglect of the forests and by the want of coal, the jury were glad to award a medal to And. Gregorini of Bergamo for his successful introduction of peat in the puddling of both iron and steel."

This was accomplished, as we have occasion to know, by the Siemens Gas Furnace, a description of which will be found on subsequent pages, under the head of "Peat in the Manufacture of Iron."

"The absence of sufficient quantities of suitable fuel, and perhaps the tendency of the national tastes, keep the manufacture of iron in the Italian Peninsula at a lower point than we should expect, who remember that some of the oldest and grandest iron mines of the world are those of Elba.

"Elba produces, on the average, forty-eight thousand tons of ore, of which twenty thousand tons are smelted in Fullonica and other places in Tuscany, and yield the excellent pigs and bars exhibited by the government.

"The total amount of pig-iron produced in Italy does not exceed thirty-eight thousand tons annually."

Falkland Islands.

These islands are destitute of coal or wood of any kind; but the lack of fuel is abundantly supplied by their extensive fields of peat, which are found in every part of the group. The deposits vary in depth from two to four feet; and it is gathered, cured, and stored at small cost.

Newfoundland.

Large quantities of peat, suitable for fuel, exist on this island.

It is stated on good authority, that beneath its surface occur the trunks and roots of trees much larger than any which are now growing on the island.

Operations for the manufacture of fuel were commenced at St. John's during the last season, but we are as yet without details of the results.

Nova Scotia.

Peat swamps and bogs are very numerous in Nova Scotia, especially in the rocky districts of the Atlantic coast. The most extensive are said to be near the Clyde River, in Shelburne, and the Carriboo Bog of

Aylesford. They consist of vegetable matter which has grown and accumulated on the spot, forming a black, carbonaceous moss, some of which has entered on the first stage of the changes by which it may be converted into coal; and it is not unusual to find in the bottom of such bogs a substance much resembling ordinary bituminous coal.

On the north-west arm of the River of Inhabitants, appears, under twenty feet of bowlder clay, a hardened bed of peat. It rests upon gray clay similar to that which often underlies peat-beds.

Pressure has rendered this peat nearly as hard as coal, though it is somewhat tougher and more earthy than good coal. It has a glossy appearance when rubbed or scratched with a knife, burns with considerable flame, and approaches in its character to the brown coals or more imperfect varieties of bituminous coal. It contains many small roots and branches, apparently of coniferous trees allied to the spruces.

ASIA.

The following, which we cut from the "New York Evening Post" of a recent date, is the only mention we have seen of anything of the character of peat in Asia, and is interesting in this connection: —

"At a recent meeting of the Asiatic Society of Bengal, Major Risely described a combustible mud, of which he first heard last September. It exists in large tracts, notably in the Pertabghur districts in Oude, where there is a jheel, or swamp of black mud, which looks like ashes, and smoulders like wood. The mud, when dried, blazes quite freely. It has been tried at Cawnpore by Mr. Taylor, the locomotive foreman, and was found to give very nearly as much steam as wood. 'It would do very well for locomotives, and could be supplied at six annas the maund.' When charred, it can be used in a blacksmith's furnace. The ash, of which it leaves a great deal, will, they say, be very useful as a manure for poor, sandy soils. Bits of bone and fragments of decayed wood were found in it at considerable depths. The Calcutta analyzers call it impure peat, resulting from the continual decomposition of vegetable matter at the bottom of a marsh. It is curious that the natives, though well aware of its properties, make no use of it, their reason being that it owes its origin to 'enormous sacrifices of ghee and grain burnt *in situ* by godlike people in old time.'"

CANADA.

Numerous and extensive deposits of peat are met with in various parts of Eastern Canada, which seems to present conditions of soil and climate peculiarly favorable to its growth and accumulation. The peat-bogs, so far as known, are chiefly confined to the plains along the St. Lawrence and its tributaries, and appear to have been formed in shallow lakes, which have been gradually filled up by the growth and decomposition of mosses. The peat often rests upon a layer of shell marl, which at one time formed the bottom of the lake. From the recent elaborate Report on the Geological Survey of Canada, made by Sir William E. Logan, Alexander Murray, Esq., T. Sterry Hunt, and E. Billings, we are enabled to mention the principal deposits of peat which are as yet known in Canada.

It is to be remarked that few of these deposits have ever yet been worked; and that it is only in a few localities that the thickness of the peat has been determined by pits or by borings.

Beginning to the westward, a deposit of peat occurs on the twelfth lot of the fourth and fifth ranges of Sheffield, where it overlies a bed of marl, and extends over three or four hundred acres. The average thickness of the peat is about four feet, aud it is said to be of a superior quality. In the level region between the St. Lawrence and Ottawa Rivers, several large peat-bogs occur; but, from their nature, the vicinity has been avoided by settlers, and they are, therefore, difficult of access. There is said to be a considerable area of peat in the rear of the seigniories of Vaudreuil and Rigaud; and also in Caledonia, where its thickness does not appear to exceed three or four feet. Peat occurs at the sources of the Pain River in Roxburg, Osnabruck, and Finch; and also in Clarence, Cumberland, and Gloucester. In the third, fourth, and fifth ranges of the latter township is a tract known as the Mer Bleu, which consists of two long peat-bogs, separated by a narrow ridge of higher land, and occupying each about two thousand five hundred acres. These deposits were sounded in many places, with a rod, to a depth of twenty-one feet, without finding bottom: in other parts, the peat was from eight to fifteen feet in thickness. This tract is situated only three miles from the Ottawa, and is about two hundred and eighty feet above the level of the sea. Three large areas of peat, of from one hundred to three hundred acres each, occur in Nepean and Goulbourn; one of them to the east, and two to the west, of the village of Richmond. It is also found on the third and eighth ranges of Beckwick, to the east of Mississippi Lake; and an area of about three thousand acres of peat occurs in Westmeath, in the rear of front A, and from the first to the fifth range behind it. In the ninth and tenth ranges of Huntley, there are about two thousand five hundred acres of peat, which in some parts has a thickness of eight or ten feet, while in other parts no bottom is found at a depth of fifteen feet. It is probable that peat may be met with in many other localities throughout this region.

On the north side of the Ottawa, three small areas of peat have been observed in Grenville. One of these, on the fourth and fifth lots, covers about thirty-six acres, and has a depth of ten feet. It has been used in the neighborhood, and is pronounced of excellent quality. Another deposit of about the same extent occurs on the first lot of the same range, and is in some parts more than fifteen feet in thickness. A third, of about thirty acres, occurs on the fourth lot of the seventh range. On the fourth and fifth lots of the first range of Harrington is a bog of about forty acres, the peat of which varies in depth from ten to twenty-five feet. Another bog is described as occurring on the first and second lots of the fifth range of the same township. It extends over about sixty acres, and has a thickness, in some parts, of twenty-five feet. All of these areas might be drained without much difficulty. To the eastward of this, a peat-bog is met with in the Rang Double of Mille-Iles. It exhibits a breadth, on the road from St. Janvier to St. Jerome, of about half a mile, and has an area of perhaps five eighths of a square mile. Its depth along the road was found to be, in several places, from two to eighteen feet, the greater depth being towards the south-east side; and its average may be taken at eight feet. A smaller deposit of peat occurs half a mile nearer to St. Janvier: it has a breadth of about a quarter of a mile, but its superficies and depth have not been ascertained. Upon the same great plain with these, a little to the north of the Church of St. Anne des Plaines, and on the north-east side of the road leading to New Glasgow, is a peat-bog, having an area of about a square mile. Its depth was not determined; but it is supposed to average about five feet. The farmers are in the habit of burning the surface of parts of this bog, and employing the ashes as a manure for the underlying portions, until by repeated burnings they reach the subjacent clay, which, mingled with the last thin layer of peat and a portion of the ash, constitutes a very fruitful soil.

Near the front of the seigniories of Assumption and St. Sulpice, there is a peat-bog, three and a half miles in length, with an average breadth of half a mile, giving an area of about eleven hundred acres. Its depth varies from two to fifteen feet; and the result of ten trials, made in two lines across the bog, gave an average of ten feet. In the seigniories of Lavaltrie and Lanoraye, there are two extensive peat-bogs, running parallel with each other. Of these the northern is the larger, and is known as the Grande Savanne. It has a length of about eight miles from north-east to south-west, and a breadth of from half a mile to two miles and a half, covering a superficies of from twelve to fifteen square miles. Two sections were made across this bog; one on the line of the railway between Lanoraye and Industry, which traverses it about three miles from its south-west extremity. It here reaches to within four miles of the St. Lawrence, and has a breadth of two and a half miles. The depth along this line was found to be from four to fourteen feet, the average of twelve trials giving about eleven feet. The other section, along the Lavaltrie Road, about four miles to the north-east, gave a breadth of half a mile, and a depth of from seven to fourteen feet; averaging, as before, eleven feet. The smaller of these bogs lies between that just described and the St. Lawrence, at a distance from the last of about two miles. On the line of the railway it has a breadth of over half a mile, and an average thickness of about five feet. It has a length of more than five miles, extending four and a half miles to the south-west of the railway, and a superficies of about three square miles.

In the fief of St. Etienne, about a mile and three quarters south-west of the Grès, on the St. Maurice River, the main road crosses a peat-bog, which is there half a mile in breadth, with an average depth of about six feet. Its extent to the north-east and south-west has not been ascertained. Another was met with in the seigniory of Champlain, about three miles from the St. Lawrence, and on the road from the church to the River Champlain. Its breadth on the road is about three quarters of a mile; and its average depth, in this part, five feet. Its length from north-east to south-west appears to be about two miles, giving to the bog an area of about a mile and three quarters. In the fief D'Auteuil, on the road between Cap Santé and the village of L'Enfant Jésus, there is a peat-bog, with a breadth of about a quarter of a mile, which has not been farther examined. Several other peat-bogs are known to exist between this last locality and the vicinity of Quebec.

On the south side of the St. Lawrence, there is a large area occupied by peat on the west side of the River Richelieu. It covers portions of the seigniories De Léry and Lacolle, and of the townships of Sherrington and Hemmingford, embracing perhaps fifteen or twenty square miles. This area is drained in part by the Lacolle River. It has not been carefully examined as yet; but it contains in some parts, particularly, it is said, in Sherrington, a very great thickness of peat. Of two specimens from this township, one, which was dark colored, fine grained, compact, and so heavy as to sink in water, gave only 3.53 per cent. of ash; while the lighter peat, from near the surface of the bog, yielded 4.66 per cent. of ash. Both of these are very pure; and the compact peat, which is remarkable from its great density and its freedom from earthy matters is particularly worthy of attention.

A large peat-bog occurs in the seigniory of Longueuil, on the road to Chambly; and an attempt was made, a few years since, to raise the peat, and introduce it to the Montreal market. A peat-bog of large size is found in the seigniory of Ste. Marie de Monnoir; and another in the parish of St. Dominique, including parts of Ste. Rosalie and St. Pie. Its dimensions may be five or six miles in one direction by three or four in another. This extent is covered by a layer of peat, which, from two or three feet at the edges, attains a depth of six feet, and in some parts, it is said, is eighteen feet in thickness. The bog has been partially drained, and portions of the land reclaimed for agricultural purposes. The drained land, being first cleared of trees, is ploughed, and then, in the dry season, set on fire. In this way, eight or ten inches of peat are burned, leaving an ash which serves as manure, and enables the surface to yield one or two crops of barley or oats. After two years, the soil becomes exhausted, and it requires to be again burned over to render it productive. When, by several repetitions of the process, the peat has been reduced to a few inches, the remaining portion is mingled, by ploughing, with the underlying clay, and a rich, mellow soil is obtained. The peat from this bog yields, when heated in close vessels, about thirty-six per cent. of coke, and contains from six to seven per cent. of ash.

In the seigniory of Rivière Ouelle, there is a peat-bog which covers about four thousand acres; and another one occurs in the seigniory of Rivière du Loup, having a superficies of six thousand acres. Its breadth, on the Temiscouata Road, is a mile and a quarter; and its depth in some parts has been ascertained to be eighteen feet. Peat is found in abundance on the first and second concessions of the seigniory of Ile Verte; and, from a point two miles below the Rimouski, there is a belt of peat-bog extending nearly all the way to the Métis River, a length of over twenty miles. Its distance from the St. Lawrence is from a quarter to half a mile, and its breadth from a quarter of a mile to a mile. The depth of the deposit, where observed, was from one to six feet. To the east of the Rimouski River, there is a peat-bog, which has a length of three or four miles, in the townships of Duquesne and Macpes, with a breadth of about three quarters of a mile, and a thickness which was found to be from five to twelve feet: it is said to be in one place thirty feet in depth. Another locality of peat is stated to be in the townships of Matanne and Macnider, between the Rivers Blanche and Matanne. A peat-bog of about one hundred acres occurs on the left bank of the Madawaska, just above the twelfth mile-post on the road to the Little Falls.

The most extensive peat deposits in Canada are found on Anticosti. Along the low lands on the south coast of the island, from Heath Point to within eight or nine miles of South-west Point, a continuous plain, covered with peat, extends for upwards of eighty miles, with an average breadth of two miles; thus giving a superficies of more than one hundred and sixty square miles. The thickness of the peat, as observed on the coast, was from three to ten feet; and it appears to be of an excellent quality. The height of this plain may be, on an average, fifteen feet above high-water mark; and it could be easily drained and worked. Between South-west Point and the west end of the island, there are many smaller peat-bogs, varying in superficies from one hundred to one thousand acres. In the vicinity of Port Daniel, in the Gaspé District, extensive deposits of peat are found.

The vegetation consists, for the most part, of mosses belonging to the genus *Sphagnum*. Besides these, however, the bogs often support a growth of tamarack and of various ericaceous plants. The leaves, roots, and stems of these, help, with the moss, to make up the peat. The peat near the surface consists of the moss but little altered, and is very soft and porous; but in the older and deeper portions of the deposit it is more dense, and darker in color, the vegetable tissue having undergone a partial decay, by which its fibrous structure, to a greater or less degree, disappears, and the substance assumes a more cheese-like texture.

The various shell marls which are often associated with the peat deposits in Canada, have, in all cases where they have been examined, proved to be nearly pure carbonate of lime.

At Bulstrode, on the Three Rivers and Arthabaska branch of the Grand Trunk Railway, Mr. Hodges has, for some three years, been conducting operations, experimental and practical, on a somewhat extensive scale, the results of which, obtained during the past year, have proved highly satisfactory.

Numerous experiments with fuel prepared at his works have been tried on regular trains over the Grand

Trunk Railway (some of which are reported on subsequent pages), from which, it is stated that "it would appear that in heating power, a ton of peat is equal to five sixths of a ton of coal, or to a cord and a quarter of wood."

These experiments having all proved so satisfactory, as to result in a contract extending over five years or seasons; during the first of which the Company are to take 100 tons per day, and during the four succeeding seasons, 300 tons per day.

The following extracts from an article which recently appeared in the "Trade Review," published at Montreal, may serve to indicate the importance which attaches to this enterprise in Canada: —

"One of the most important questions that can arise for a country like Canada, where winter reigns supreme for from four to six months out of the twelve, is that of affording a cheap means of providing the heat actually necessary to existence.

"We have hardly any data on which to base a calculation of the yearly cost to the country of the fuel consumed, but we may approximate to its cost in some measure.

"Making due allowance, on the one hand, for these fortunate localities, where cord-wood can be had for the cutting and hauling, and where the cost may be set as low as two dollars per cord, and also taking into consideration, on the other hand, the large quantities of wood (or coal at an equivalent valuation) consumed in cities at six, seven, and even eight dollars per cord, we think the average cost of fuel may be taken at four dollars and a half, which, in all probability, is below rather than above the mark. Now, let us say that there are four hundred thousand families in Canada, burning at the rate of twelve cords per annum (and this, too, is a low estimate when allowance has been made for the fuel used for steam purposes, and in warming churches, stores, warehouses, etc., etc.), and the consumption of fuel will represent a total cost of $21,600,000, or not far from double the entire expenses of carrying on the government of the country. It will, therefore, at once be seen how great is the economic importance of endeavoring to provide fuel at the lowest possible cost, as well as of using that fuel in the most profitable way.

"Every invention which will tend to secure economy, either in cost or consumption, will increase by so much the wealth of the country, in setting free, for other productive purposes, capital and labor now employed in the cutting and carrying to market of wood, and in the importation of coal.

"One description of fuel, hitherto unused, though not exactly unknown in Canada, is *peat*. Several experiments have been made in past years, and unsuccessful attempts to introduce it into general use; but, at last, it seems a mode of preparing it for market well and cheaply has been discovered; and, if only a sufficient quantity can be supplied to meet the demand, one mode of economizing fuel will have been obtained.

"By the experiments of the Grand Trunk Railway, it would appear that, in heating power, a ton of peat is equal to five sixths of a ton of coal, or to a cord and a quarter of wood. We are not aware that any careful experiments have been made to test the comparative heating and lasting qualities of peat, and the various kinds of coal and wood; and these would be necessary before any exact calculations could be made as to the gain in the use of the new fuel. But, as far as we can judge from the trials made, the gain will be at least from thirty to thirty-five per cent., as compared with wood or coal. Now, a saving of even thirty per cent. on our estimate of the annual consumption of fuel will be nearly six and a half millions of dollars, and, were the precise facts known, would amount to very much more. Were this saving to be applied to the payment of the public debt, it would not take long to reduce it to very small proportions; and it is a question whether the government should not take steps to aid private enterprise — say by opening credits in a judicious way — to secure so very great an advantage.

"In addition to the experiments to test its qualifications for steam-raising purposes, a trial has lately been made with a view of seeing how the peat would answer for smelting purposes. Experienced gentlemen present, who watched the experiment with great interest, pronounced the castings to excel in toughness and quality of chill any specimens they had before seen.

"With regard to the probabilities of a sufficient supply, peat is to be found in a great many localities, and in great abundance; and, according to the official reports of the Geological Commission, in the following places throughout the Province: Sheffield, Caledonia, Gloucester, Huntley, Grenville, Harrington, Mille-Iles, St. Anne des Plaines, St. Sulpice, Lavaltrie, St. Maurice, Champlain, Lacolle, St. Dominique, Riviere Ouelle, Riviere du Loup (en bas), Dufresne, Sherrington, Longueuil, and the Island of Anticosti.

"In the last mentioned locality, the beds are very extensive. One of them possesses a superficies of not less than one hundred and sixty miles, and several others an extent of 4,000, 6,000 and 10,000 square acres respectively."

The "Montreal News," alluding to the efforts now being made to utilize peat, says, —

"There is no social question that causes more anxiety to those friends of Canada who peer into the future than the difficulties of securing a supply of cheap fuel. Great suffering and privations are even now endured in old settlements in consequence of the destruction of the forests. In the houses of many a habitant who owns a good farm, the fires are put out at nine o'clock to economize wood; roots are dug up, and branches gathered for fuel, which in by-gone years would have been rejected. Yet year by year the forest is falling back, and the price of an indispensable article of consumption rising in price."

The supply of material from which to produce a superior fuel at small cost is nowhere more abundant than in Canada, and it will require only a moderate amount of enterprise and capital to place it in the market.

Maine.

Dr. C. T. Jackson, in his geological survey of this State, directed attention to the numerous valuable deposits, or rather accumulations, of peat.

At the localities which he designates, this substance may be most advantageously wrought for fuel. It is said by him to be found at the depth of three feet from the surface, amid the remains of rotten logs and beaver sticks; showing that it belongs to the recent epoch. Some of the deposits are twenty feet in thickness, resting on white silicious sand.

In one instance, coal is stated to have been found while digging a ditch to drain a bog. He says, "On examination, I found that it was formed from the bark of some tree allied to the American fir, the structure of which may be readily discovered by polishing sections of the coal, so that they may be examined by the microscope." His analysis shows that it contains —

Bitumen	72
Carbon	21
Oxide of Iron	4
Silica	1
Ox. Manganese	2
	100

"This substance is, therefore, a true bituminous coal, remarkable indeed for containing more bitumen than is found in any other coal known. I suppose it to have been formed by the chemical changes supervening upon fir balsam during its long immersion in the humid peat."

This is a very interesting discovery; and the same substance appears to exist in other peat-bogs of the State.

The same author notices the accumulation of peat at Quoddy Head, near the south-east angle of the State. This deposit is fifteen feet in thickness, and is of an excellent quality for fuel.

"The time may arrive, when, even in Maine, wood becoming scarce, her neglected peat-bogs will be resorted to for fuel; though here, as in many other sections, were the *superiority* of the article over even wood or coal known and appreciated, the bogs would be worked now rather than to await the period at which, for lack of other fuel, their valuable deposits shall be drawn upon."

Dr. Jackson adds, "There are so many localities of peat in Maine, that I have thought it hardly necessary to describe them, but would, however, point out localities on the railroad route in Bangor, at Bluehill, near the Marsh Quarry in Thomaston, in the town of Limerick, and on the Coolidge Farm in Waterford. Near Lewiston also are very considerable deposits, which have been worked to some extent. These localities are among the most abundant, and may be most advantageously wrought for fuel, which may be used for the burning of lime and for domestic purposes." Valuable deposits are said to exist also in the vicinity of Portland, Augusta, Kennebunkport, and Bangor, and we learn of practical operations proposed in several of these localities for the coming season.

The Island of Campo Bello, at the most easterly extremity of the State, and now celebrated as the spot where Fenian squadrons were first "set in the field" during the late attempt to capture British North America, is said to have been recently sold for its "mineral resources;" but it has since transpired that a very large proportion of the area is *peat*, of excellent and uniform quality, and it would seem to be a pertinent question, whether a small amount of capital invested in manufacturing this material into fuel would not be likely to bring larger and quicker returns than a very large sum expended in developing the "mineral resources."

New Hampshire.

In this State, peat has been used but very little as an article of fuel; though, as an important ingredient in forming a compost, it is extensively used, and its value recognized. Deposits are numerous, and in some places are known to be of excellent quality for fuel; as in Meredith, Canterbury, Rochester, and Franconia, especially the three former. Valuable deposits are also found in Northumberland and vicinity, Whitefield, Littleton, Dublin, Warner, Franklin, Kingston, Lancaster, Lyndeborough, and other localities.

Extensive deposits, some of them of superior quality, are found in Rockingham County; and it is understood that the manufacture of fuel is to be undertaken on a large scale, in one or more localities within its limits, the coming season.

Vermont.

The Report on the Geology of Vermont, by Professor Hitchcock and Mr. A. D. Hagar, contains many items of interest concerning peat, from which we extract most of what follows.

Peat-beds may be found in every town in the State. The number is so great that we shall not endeavor to enumerate them all. We will, however, mention some of special interest, mostly from the observations of Rev. S. R. Hall.

The peat throughout the State, it is thought, will average about five feet in thickness, and is in many places covered with a growth of timber consisting of cedar, black ash, tamarack, spruce, and pine. Occasionally it is reported of great depth, as in the following, which is an extreme case: —

In boring for water near Whiting Depot, on Otter Creek flats, ten miles south from Middlebury, they pierced through eighty feet of peat, and at seventy-two feet from the surface passed a sound log of wood. The land is now covered with a heavy growth of pine and ash, while under the roots of the standing timber are distinctly visible the decaying trunks of a former growth. How old is that bed of peat?

In Albany, on the farm of Zuar Rowell, east of Great

Hosmer's Pond, overlaying marl, is a bed of peat four feet thick, covering from six to ten acres. It is computed that each acre contains about one thousand five hundred cords of peat.

On the land of G. W. Powers, in the south-east part of the town; half a mile north, upon Mr. Orne's farm; in the north-east part of the town, on land of Mr. Church, Mr. Hovey, and others, there is a large amount of peat. Near the new church there is a bed of peat, evolving sulphuretted hydrogen when disturbed.

Peat is also described as abundant in the towns of Andover, Barnet, Barre, Barton, Berlin, Bethel, Bradford, Brattleborough, Bridgewater, Brookfield, Brownington, Calais, Cavendish, Chelsea, Chester, Corinth, Coventry, Craftsbury, Danville, Derby, Dummerston, Eden, Elmore, Glorr, Greensborough, Hancock, Hardwick, Hartford, Holland, Hydepark, Jamaica, Londonderry, Lowell, Ludlow, Marshfield, Montpelier, Moretown, Morristown, Newfane, Norwich, Peacham, Plymouth, Putney, Pownal, Randolph, Rochester, Rockingham, Royalton, Ryegate, St. Johnsbury, Springfield, Thetford, Townshend, Troy, Waitsfield, Walden, Wardsborough, Warren, Washington, Waterbury, Waterford, Weathersfield, Westminster, Westmore, Williamstown, Windsor, Woodbury, Woodstock, Wolcott, and Worcester. If particulars concerning these beds are desired, they will be found described as *muck*, in the Second Geological Report of Professor Adams.

Numerous, and in some cases very valuable, beds of marl are disposed in all parts of the State; and the remark is made, that most of them are associated with beds of peat and other forms of organic vegetable matter.

Many of the peat-beds are observed to occur in

meadows formerly the haunts of beaver; and numerous traces of the works of these busy little animals are discovered as excavations are made.

It is an interesting fact, that one of the most remarkable fossils ever found in New England was discovered in a *peat-bed* in Vermont, on the line of the Rutland and Burlington Railroad.

We allude to the remains of an elephant, which were found in Mount Holly in 1848. The following is Professor Thompson's description of it: —

"The Rutland and Burlington Railroad crosses the Green Mountains, in the township of Mount Holly, at an elevation of one thousand four hundred and fifteen feet above the level of the ocean; and the bones of the fossil elephant were found at that height. It was in a *peat-bed*, east of the summit station, that these bones were found. The basin in which the peat is situated appears to have been originally filled with water, and to have been a resort for beaver, a large proportion of the materials which formed the lower part of the peat consisting of billets of wood about eighteen inches long, which had been cut off at both ends, drawn into the water, and divested of the bark by the beaver, for food. The peat was fifteen feet deep before the excavation for the railroad was made.

"In making this excavation, the workmen found, at the bottom of the bed, resting upon gravel, which separated it from the rock below, a huge tooth. The depth of the peat at that place was eleven feet. Soon afterwards, one of the tusks was found, about eighty feet from the place of the tooth above mentioned, which was a grinder. Subsequently, the other tusk and several of the other bones of the animal were found near the same place. These bones and teeth were submitted to the inspection of Professor Agassiz, who pronounced them to be an extinct species of elephant.

"The grinder is in an excellent state of preservation, and weighed eight pounds; and the length of its grinding surface is about eight inches. The tusks are somewhat decayed, and one of them badly broken. A cord drawn in a straight line from the base to the point of the most perfect tusk measures sixty inches; and the longest perpendicular, let fall from that to the inner curve of the tusk, measures nineteen inches. The length of the tusk, measured along the curve on the outer surface, is eighty inches, and its greatest circumference twelve inches. The circumference has diminished very much since the tusk was taken from the peat-bed, on account of shrinkage in drying; and several longitudinal cracks have been found in it, extending through its whole length; and it was found necessary to wind it with wire to prevent it from splitting to pieces."

In 1858, the remains of another elephant were found in Richmond, which are now in the cabinet of Vermont University.

So late as 2d Sept., 1865, the tusk of a fossil elephant was found in a *peat-bed*, about five feet below the surface, on the farm of D. S. Pratt, in Brattleborough, by a man-who was excavating for compost. The following account of it is given in "The Vermont Record" of 5th Sept. The tusk is forty-four inches in length, and eighteen inches in circumference at the largest end, and eleven inches at the smallest. It is in a fair state of preservation, although some parts of it crumbled after being exposed to the air. The workman, on discovering it, took a piece to Mr. Pratt, remarking, as he handed it to him, that he had found a curious piece of wood.

Mr. Pratt, on looking at it, discovered its true nature. This tusk belonged to a species of elephant long since extinct, supposed to be the *Elephas Primogenius* (or mammoth), Blumenbach, that inhabited the northern parts of North America, having wandered across the Siberian plains to the Arctic Ocean and Behring Straits, and beyond to this country south to about the parallel of forty degrees. Their bones show them to have been about twice the weight and one third taller than our modern species."

Rev. S. R. Hall, in his Report on the Geology of Vermont, says, "Nearly all the peat or muck of Vermont would answer a good purpose for fuel; but at present it is not needed. It would furnish an abundance of carburetted hydrogen, if employed for producing *gas-light*, much less expensive than coal, oil, or resin. The gas is harmless, inoffensive; and has, in respect to healthfulness, great advantages over some other kinds."

On the summit of Mansfield Mountain, in the town of Stowe, the highest point in the State, 4348 feet, are found beds of peat and the *sphagnous moss* that produces them. This is one of the very few cases in New England where peat occurs at great elevation. In Europe, the cases are numerous.

Massachusetts.

In his Report on the Geology of Massachusetts, Dr. Hitchcock says, "Taking the State as a whole, peat is but little used, either as fuel or manure; yet for both purposes its use is rapidly increasing, especially in the eastern part of the State, where fuel is more expensive. In view of its importance, I have made some efforts to ascertain its probable amount in our swamps. But this is very difficult; both because our swamps where it occurs have been but slightly explored, and because much is called merely mud that deserves the name of peat."

He then proceeds to enumerate various localities where it is known to exist, and gives data for forming an approximate estimate of the amount of this deposit in the State, and adds, "It will be seen that scarcely any towns in the four western counties of the State are mentioned. This is partly explained by the fact that fuel is more plenty there than in the eastern counties, so that public attention has never been directed so much to our fossil resources. But I think it undeniable that the amount of good peat in the western counties is much less than in the eastern.

"Although, perhaps, the swamps abound as much in vegetable matter that would be useful in agriculture, yet it does not seem to be converted into genuine peat; though I do not doubt that it will be easy to find a large amount of it when there is a demand for it.

"Excluding these western counties, and taking the amount of peat given in the statements made to me as a fair average of its quantity in all the towns of the other counties (excluding the large towns), it would follow that eighty thousand acres, or one hundred and twenty-five square miles, are covered with peat in that portion of the State, having an average thickness of six feet four inches. This area and depth would yield not far from one hundred and twenty-one million of cords.

"If this should be thought by any to exceed the quantity of good peat existing in that section, I presume no one will consider it too high an estimate of the amount of swamps filled with vegetable matter. I presume it falls far short of the true amount.

"We hence get an enlarged view of the quantity of matter in the State that may be employed as fuel, or in agriculture, which has hitherto, except in some limited districts, remained almost untouched.

"It is true that peat (unmanufactured) is not so convenient and agreeable a kind of fuel as good wood or coal; yet it certainly answers a very good purpose; and the facts in the case tend to allay the apprehension which must sometimes rise in the mind of one who sees in the gradual diminution of our forests a future check to our prosperity and population.

"It is gratifying to learn from so many towns that the inhabitants are awaking so much to the use of peat and peaty matter. Some gentlemen have even spoken of it as a 'peat fever.' I hope it has not yet reached its crisis."

Recent personal experience, and considerable intercourse with numerous persons who have during the past season used *solidified* peat as fuel, has convinced us that Dr. Hitchcock underestimated the convenience and agreeableness of peat as compared with "good wood or coal." We consider it preferable by far to either, in many respects. His remarks referred to the crude, unmanufactured article.

The following list of towns where peat is known to exist includes those mentioned by Dr. Hitchcock, and others of which we have personal knowledge: —

Abington,	Athol,	Bernardston,
Acton,	Barnstable,	Billerica,
Amesbury,	Bedford,	Boylston,
Andover,	Bellingham,	Brewster,

Bridgewater,
Buckland,
Cambridge,
Carver,
Chelmsford,
Chilmark,
Cohasset,
Concord,
Danvers,
Dennis,
Dighton,
Dover,
Duxbury,
Eastham,
Falmouth,
Farmingham,
Greenfield,
Groton,
Hadley,
Halifax,
Hamilton,
Hanover,
Hanson,
Hingham,
Holden,
Hopkinton,
Ipswich,
Kingston,
Lancaster,
Leverett,
Lexington,
Longmeadow,
Ludlow,
Lunenburg,
Lynnfield,
Medfield,
Medway,
Methuen,
Millbury,
Milton,
Nantucket,
Natick,
Needham,
Newton,
Northborough,
Orleans,
Oxford,
Pittsfield,
Randolph,
Reading,
Rowley,
Roxbury,
Seekonk,
Southborough,
South Reading,
Sharon,
Shrewsbury,
Shutesbury,
Spencer,
Stoughton,
Sudbury,
Sunderland,
Tisbury,
Topsfield,
Truro,
Uxbridge,
Wales & Holland,
Walpole,
Waltham,
Watertown,
Wellfleet,
Westford,
Weston,
Wilmington,
Wrentham,
Yarmouth.

In his "Geology of Massachusetts," 1833, Dr. Hitchcock says of peat, "There are two varieties, the fibrous and the compact. In the former, the moss, turf, and roots, out of which peat is formed, have not lost their fibrous structure; but in the latter they are converted into a compact and nearly homogeneous mass.

"The fibrous and compact varieties probably exist at nearly every locality.

"I am sure of their occurrence in Cambridge, Newton, and Lexington, and in large quantities. Peat is abundant in Seekonk, Uxbridge, Cohasset, Duxbury, Hingham, Medfield, Walpole, Wrentham, Dover, Framingham, Sudbury, Topsfield, Ipswich, and Nantucket.

"It exists and has been dug, in greater or less quantities, in Pittsfield, Hadley, Leverett, Shrewsbury, Lancaster, Southborough, Hopkinton, Medway, Halifax, Stoughton, Boylston, Reading, Milton, Needham, Concord, Billerica, Bedford, Waltham, Watertown, Acton, Wilmington, Danvers, Chelmsford, Hamilton, and in nearly all the towns in Barnstable County; certainly in Yarmouth, Brewster, Orleans, Eastham, Wellfleet, and Truro.

"I cannot but regard the existence of so large quantities of peat on Cape Cod and Nantucket as a great blessing to the inhabitants; yet, from the little of it which I observe to be dug there, I am apprehensive they do not realize its value.

"Not a town in the State can be named where more or less peat does not exist. The eastern section, however, is certainly best stored with those varieties that may be employed for fuel; and it is an unexpected fact, that the south-east part of the State, which abounds with sand, contains also a large amount of peat.

"The beds of peat on Nantucket, and the small adjacent Island Thuckanuck, Muskegut, and Gravel, contain, according to a report of Lieutenant Prescott, six hundred and fifty acres of peat from one to fourteen feet thick, and generally of good quality. This must afford an inexhaustible supply of fuel for the inhabitants; and yet I was surprised to learn that so little use was made of it."

We are told of a deposit of peat at Barnstable, resting upon a thick bed or matting of cranberry vines, the vines and berries still preserving their original form until brought to the surface and exposed to the sun and air.

We have ourselves seen at Lexington, Mass., a rich deposit of pure black peat, six feet in depth, resting upon a compact bed of clean moss, of the consistency of wet hay, and retaining its light green color.

It is a curious fact also, and worthy of note, that, along the coast in the south-east part of the State, the remains of ancient forests, now submerged, are not uncommon.

This is the case in the harbor of Nantucket, as reported by Lieutenant Jona. Prescott, who superintended the dredging of that harbor. Portions of cedar, maple, oak, and beach trees were found, some of them in an erect position, accompanied by peat. Another submarine forest exists at Holmes Hole. Near the southwest extremity of the Vineyard, we learn of another. On the north side of Cape Cod also, opposite Yarmouth, cedar stumps are found, extending more than three miles into Barnstable Bay. The same thing is said to occur in the bay of Provincetown, on the side opposite the village.

Professor Lyell, in his second visit to the United States, mentions a submerged forest "at Hampton, on the way from Boston to Portsmouth;' also one near Portsmouth, N. H., "now submerged at low water, containing the roots and upright stools of the white cedar, showing that an ancient forest must once have extended farther seaward." In his first visit to North

America he mentions a submerged forest, somewhat similar, near Fort Cumberland, in Nova Scotia. He records other observations in relation to submerged trees at the mouth of Cooper River, near Charleston, S. C., and of the Altamaha, in Georgia.

A letter from Elias Phinney, Esq., of Lexington, dated Jan. 30, 1839, and addressed to Dr. C. T. Jackson, upon the subject of peat, has been many times referred to, and so often quoted that it may be regarded as authority. It will be found copied entire in the Report of the New York Survey, submitted by Dr. W. W. Mather, in December, 1839; and in the Geological and Agricultural Report of Rhode Island, 1840. A few extracts from it will be of interest here: "I consider my peat-grounds by far the most valuable part of my farm; more valuable than my wood-lots for fuel, and more than double the value of an equal number of acres of my uplands for purposes of cultivation.

"In the first place, they are valuable for fuel. I have, for twenty years past, resorted to my peat-meadows for fuel. It gives a summer-like atmosphere, and lights a room better than a wood-fire. The smoke from peat has no irritating effect upon the eyes, and does not, in the slightest degree, obstruct respiration, like the smoke of wood; and it has none of that drying, unpleasant effect of a coal-fire. The ashes of peat are not more troublesome, and are less injurious to the furniture of a room, than the ashes of coal.

"The best peat is found in meadows which have for many years been destitute of trees and brush, and well drained, and where the surface has become so dry, and the accumulation of decayed vegetable matter so great, that but little grass or herbage of any description is seen upon the surface.

"A rod square, cut two spittings deep, each spitting of the length of eighteen inches, will give three cords when dried. It may be cut from May to September. If the weather in autumn be very dry, the best time for cutting will be from the middle of August to the middle of September. If cut in the latter part of summer or early in autumn, it dries more gradually, and is not so liable to crack and crumble as when cut early in summer. It is considered a day's work for a man, a boy, and a horse, to cut out and spread a rod square. It will require about four weeks of dry weather to render it fit to be housed for use.

"Peat taken from land which has been many years drained, when dried, is nearly as heavy as oak-wood, and bears about the same price in the market."

As early as 1815, the subject of peat as an article of fuel was by some considered of so much importance, that Dr. Aaron Dexter, President of the Massachusetts Agricultural Society, submitted to the trustees a paper which he had prepared upon the subject, being an abridgment of a work in two volumes, by Dr. Robert Rennie, of Kilsyth, in Scotland, accompanied by some remarks upon the valuable deposits in our own State, and the importance of improving them. The paper was published; and from it we make the following disconnected extracts: —

"Peat-water possesses astringent, antiseptic qualities. Rain or river water, when allowed to stagnate, especially in warm weather, becomes putrid: peat-water does not.

"Rosier observes that the air of peat-mosses is always salubrious; that, by a wonderful provision of Nature, oxygen is exhaled, and hydrogen absorbed; and, by reason of the low temperature of the moss, carbonic acid is not ėvolved, so that by this means the air is never infected with these deleterious gases.

"Carr, in his 'Stranger in Ireland,' mentions that this is the case in that country. Dr. Walker makes a similar remark. He says, 'Stagnant rain-water, especially in warm weather, occasions, in fenny countries, intermittent fevers and putrid diseases, whereas no such effects are felt from stagnant moss-water.' The moors and mosses in Scotland, and the turf-bogs in Ireland, are inhabited by as healthy people as any in the world. No intermittent fevers, no putrid sore throats, prevail among them. M. de Luc makes the same remark. He says, 'Over the whole continent of Europe, which I examined, the air of the mosses, even in the lowest valleys, is very salubrious. The inhabitants of these districts are remarkably healthy: they are not liable to the fever and ague which prevails in other low level lands in their immediate vicinity, where there are no mosses.' The oxygen they discharge must purify the air, and promote the health of those who inhabit such districts.

"Some peat, when dug, becomes so hard, so heavy, and so glossy in the fracture, that it is with difficulty it can be distinguished from some of the softer coals. On the contrary, Mr. Williams says that he has seen coal so soft, that it was with difficulty it could be distinguished from the hard, black, glossy peat.

"There are the strongest reasons for supposing that all coal has been at one period of its formation in a soft, pulpy state, like the above species of peat when newly dug; that coal, wherever it has been discovered, has certainly been exposed to a degree of mechanical pressure far beyond that which was ever applied to peat by art. It would be superfluous to offer any proof of this; and, if the best peat was subjected to the same degree of compression, it is obvious that it would become equally compact and equally heavy, bulk for bulk, and equally as inflammable as coal; in no respect distinguishable from that substance, in color, consistency, or chemical qualities. The external appearance of coal shows its alliance to peat in color and consistency: the resemblance is often clear."

Rhode Island.

Dr. Jackson, in his Report on the Geological and Agricultural Survey of this State, says, "Peat and swamp muck occur in almost every town in the State;" and goes on at considerable length to give his views in regard to the value of both, the former as fuel, and both as fertilizers.

He mentions the following localities in which peat of excellent quality for fuel is found, as shown by analysis, statements of which are incorporated in his report: —

Block Island,	North Kingston,	Warwick,
Bristol,	Nyatt,	Woonsocket,
Cranston,	Pawtuxet,	Wickford.
Cumberland,	South Kingston,	

We also learn of extensive deposits, of excellent quality, in the immediate vicinity of Providence; arrangements are being made to work some of them on a pretty large scale, and there can be little doubt that the already extensive and prosperous manufacturing interests of the State will be largely promoted by this newly-found source of fuel.

CONNECTICUT.

This State is rich in deposits of peat; and although, until recently, little use has been made of it except for agricultural purposes, her people are among the earliest to perceive the importance of putting it to practical account in the shape of fuel, and to avail themselves of the methods of manufacturing it: and it can hardly be matter of surprise now, if the little State which has so long enjoyed the reputation of making nutmegs from beech-wood should succeed even better in producing a solid fuel from her hitherto worthless and neglected bogs.

Professor S. W. Johnson, of Yale College, in an exceedingly interesting and valuable essay on peat and muck, especially as relates to their nature and agricultural uses, — which was published at Hartford some years since, but is now, as we are informed, out of print, — gave an amount of information upon the subject, which it is to be regretted should have been limited in the extent of its circulation to a single small edition.* In addition to a thorough treatment of the whole subject, as relates to the use of peat for agricultural purposes, he had been at considerable pains to obtain somewhat detailed statements concerning the character and extent of peat-beds in different parts of the State, with their several peculiarities of formation, location, elevation, surroundings, &c.; and accompa-

* The work has recently been revised and enlarged, and under the title of "Peat and its Uses as Fertilizer and Fuel, by Samuel W. Johnson, M. A., Professor of Analytical and Agricultural Chemistry, Yale College," is published by Orange Judd & Co., 41 Park Row, New York. 1866. Price, $1.25.

nied it with report of analyses of numerous specimens obtained from the bogs described.

These show clearly that there are stores of good peat within the State, although there are some meadows, which, from their peculiar situation, have been subjected to so much wash of earthy matter from the surrounding hills, that the fuel to be produced from these would necessarily be of a somewhat inferior quality.

He mentions peat-beds in Goshen, Milford, Plainville, Berlin, Griswold, Colebrook, Cornwall, Granby, Brooklyn, Poquonock, Collinsville, New Haven, Rockville, Stonington, New Canaan, and Salisbury. We might add Hartford, Bridgeport, Meriden, Somers, Broad Brook, New London, Willimantic, Waterford, and numerous other localities, of which we have had more or less personal knowledge. Certain it is, that some of the finest specimens of peat we have ever seen, and best adapted for fuel, have come from meadows in Connecticut.

Preparations, in some places quite extensive, are being made for the manufacture of fuel the coming season; and, while the demand for domestic purposes will doubtless absorb a very considerable portion of the product of the first season, the numerous manufacturing establishments and railroads, some of which are in close proximity to large beds of very pure peat, will doubtless, without much delay, discover its value and economy for their several requirements, and thereby create a demand which can be supplied only by constantly extending works.

These deposits of peat are a source of wealth to individuals and the State, the extent of which, it is presumed, few will comprehend until the developments now in progress shall have begun to demonstrate it.

New York.

In the several Geological Reports of this State, made by Professor Mather, Mr. Vanuxem, and Dr. Emmons, special mention is made of the rich deposits of peat, and the certainty, that, sooner or later, they will undoubtedly prove to be of great value for the production of fuel.

A great number of localities are indicated and estimates made of the amount of peat "on deposit."

Dr. Emmons furnishes some particulars of peat within his district, points out the high economical value which he considers must attach to it as a combustible, and says, "Perhaps it would be saying too much to assert that peat is more valuable than coal: but when we consider, that, for creating heat, it is not very inferior to bituminous coal; that it contains a gaseous matter equal in illuminating power to oil or coal-gas; that its production is equally cheap; and, in addition to this, that it is a valuable manure if properly prepared, — its real and intrinsic worth cannot fall far short of the poorer kinds of coal."

It should be borne in mind that these remarks refer to the article in its *crude*, unmanufactured state. It will, we anticipate, be found that the *manufactured and condensed* peat produced by the method recently introduced, is, for many purposes of domestic use, but especially for steam, and in the manufacture of iron and steel, superior to any other kind of fuel now in use.

This statement will undoubtedly be considered an extravagant one; but facts which are daily coming to our notice are fast conspiring to demonstrate that such is the case.

Dr. Mather, in his report in 1838, says, "During the surveys of the past season, I have collected specimens of peat from various localities. Some of them, now when dry, are compact enough to receive a slight polish, and have as great a specific gravity as bituminous coal, and would probably give out as much heat."

In his report in 1839 on New York, Westchester, and Putnam counties, he says, "Peat is now coming into use as a fuel, and must, before many years, be extensively employed for this purpose in this part of the country, where coal and wood are so expensive. The marshes of the Hudson River, in these counties, that will yield peat, may be estimated at 1000 acres, with a yield of 2000 cords per acre, or 2,000,000 cords. These include those near Sing Sing, Verplanck, Peekeville, Anthony's Nose, Constitution Island, and numerous smaller ones. The peat in some of these marshes, where it was examined, is of inferior quality, fibrous, and contains much earthy matter. That formed in marshes in the interior of those counties is of much better quality, and far superior as an article of fuel." And later, in 1840, he writes: "I would again urge upon our farmers and other citizens the importance of making use of peat for fuel. It is a cheap and valuable fuel; and, when properly prepared, it also makes one of the best renovators of the soil. Peat is equal in value to oak-wood, bulk for bulk."

Rev. Mr. Shafter, of New York, says, "The peat, by drying, acquires a high degree of solidity. It is easily kindled, burns with a bright flame, yields a bluish smoke, and produces an odor similar to that which attends the combustion of gramineous substances. But this is momentary. When thoroughly kindled, it burns with less flame, yields a small proportion of blackish smoke, and sulphurous acid gas is evolved, though I cannot discover any pyrites. It burns for a long time, and emits a great body of heat. It leaves a very small proportion of light and grayish-white ashes. I cannot refrain from expressing my opinion that this variety of peat will answer as an excellent substitute for the best Liverpool coal."

Professor E. Emmons says, "From the abundance of peat in this State, it appears that the climate and other circumstances are favorable to its production. It is not so hot as to cause a rapid decomposition of vegetable matter, nor so cold as to prevent those changes, somewhat allied to fermentation, which are required for its formation.

"Contrary to expectations, I find it in great abundance in the counties of Clinton, Warren, and Hamilton; and I may state, in general, that most of the flies in those counties abound in this substance.

"The only places where it can occur are those of a marshy character, and the substance itself may be tested by any person by first drying and then igniting it. If it burns, it is peat. As its presence may be suspected in all low, wet places, especially those bordering on ponds and lakes, it will be well to search for it in all such places by thrusting down a pole or stick, and trying the matter that adheres to it, as it regards its combustibility; or it may generally be found wherever the surface of the ground is easily agitated by passing over it. One of the largest collections of this substance, which has fallen under my observation, is in Champlain, in the county of Clinton. The peat marsh or fly to which I refer is in the west part of the town, and is about two or two and a half miles in length, and from a half to three fourths of a mile in width. Over the whole extent of this fly, a pole may be thrust down from twelve to thirty feet, and probably in many places to twice thirty feet. It is, of course, nearly inexhaustible. Others of nearly equal extent occur in the county, and many which are less extensive. One fact which applies to all the peat-marshes of this neighborhood is, that they are situated far above the level of the lake, and that those marshes which are on or near the same level as the lake do not contain peat.

"Peat, as is well known, answers a good purpose for fuel, and undoubtedly ranks next to coal for sustaining for a long time a high temperature. There is no substance which would remove so much suffering among the poor as the general introduction of this substance for fuel in our larger towns and cities. Its abundance and cheapness recommend it to the attention of the public; and if measures could be devised to bring it into use in this State, many important results would follow."

The following are some of the most valuable localities in the State, though there may be many others of equal or greater importance: —

A large portion of what was once a part of Peat Marl Pond, four miles north of Kinderhook, is filled with peat, and has become a marsh. Other parts of this pond are filling with the same combustible and with shell marl. It is estimated that sixty or seventy acres of this pond and marsh are filled with peat. It has probably a mean depth of at least six feet, and ought to yield sixty thousand cords of good peat.

A small bog, of about six acres, is on the farm of Cornelius P. Van Allen, three miles north of Kinderhook. The peat is of fine quality, at the depth of two feet from the surface. It has a mean depth of about nine feet, and ought to yield nine thousand cords of good peat.

Another, similar to the above, in the north part of Stuyvesant, contains about the same quantity.

Round Pond, in the north part of Kinderhook, contains two thousand to three thousand cords.

The marshes and shallows of Kinderhook Lake probably contain twenty thousand to thirty thousand cords.

A small bog between this lake and North Chatham contains perhaps two or three acres of peat.

In the marsh west of the post-road, one mile north from Kinderhook, there is said to occur a considerable quantity of peat.

The marsh belonging to Mr. Lucas Hoes, one mile south-west of Kinderhook, near the post-road on the east side, contains about thirty acres, with a mean depth of six feet.

Several other localities are said to occur in the valley of Kinderhook Creek, between Kinderhook Village and Stuyvesant town-line.

A peat-bog is also said to occur two miles north-east of Valatie.

Peat-bogs occur in many places in New Lebanon, among which may be mentioned those on Mr. Gillett's and the adjoining farms, and on Mr. Tilden's.

Another, south of Mr. Carpenter's, of fifteen acres, and three to twelve feet deep.

Another, south of Fitch & Kirby's store, owned by Mr. Waite, of about thirty acres.

Peat occurs near the west side of Canaan Mountain, around Adgate's Pond. The aggregate amount in this township is probably four hundred thousand cords.

Peat occurs on Rowland Story's farm, a quarter of a mile east of Lafayette Corners, in Milan.

A bog of peat, five or six acres, with a depth of five or six feet, occurs a mile and a half east of Upper Red Hook.

Another between Stormville and Hopewell.

Several bogs between Hopewell and Fishkill appear to be peat-bogs. They contain probably an aggregate surface of forty acres.

A peat-bog of five or six acres is located about a mile and a half from Stormville, on the road to Beekman.

An extensive peat-bog extends north from Long Pond down the valley of its outlet.

A large body of peat is said to exist on Mr. Legget's farm, in Ghent.

Peat probably exists in the marsh east of Great Nutten Hook.

Extensive deposits are found a mile or two west of Malden.

Two or three small peat-bogs occur in the south part of the town of Ghent. They may contain ten or fifteen acres.

A small marsh of ligneous peat occurs about a mile and a half north of Hillsdale.

There is an extensive peat-bog on Lawrence Smith's and the adjoining farms in Amenia. Professor Cassels reports it to have an area of about one hundred and fifty acres, containing probably one hundred and fifty thousand cords.

He also reports that there is a peat-bog four miles north-east of Dover, on the east side of the creek; another, one mile south of the above; another, one mile south of the last mentioned; one also two miles south of Dover; and one eight miles south of Dover. These contain an aggregate of probably seventy-five thousand cords.

Extensive peat-bogs are found east of Elbow Mountain, which lies east and north-east of Dover. They are in the valley through which the road passes from Kline Corners to the Columbia Furnace, in Kent. The northern one is about three quarters of a mile long, and two hundred to three hundred yards wide, with an unknown depth. Its depth was measured in several places, and it was generally five feet deep within five rods of its edge. It was once a lake, now filled with peat. Its mean depth may probably be placed at nine feet, its area at sixty acres, and its contents at ninety thousand cords.

The other bog south of this probably contains forty acres, with a depth of six feet; and its contents may be estimated at forty thousand cords.

Peat is found abundantly in the vicinity of Pine Plains, and some of it is of very good quality. A small bog is observed one mile south of Pine Plains. Cranberry Marsh and Cedar Swamp, near Stessing Pond, are filled with peat. It is rapidly forming in some parts of Stessing Pond.

Peat is forming on Woodward's farm, in Copake. In Taghkanic, about a mile and a half or two miles west of Crysler's Pond, a peat-bog occurs of thirty or forty acres.

From descriptions of the drowned lands in Aucrain, it is probable that peat abounds there. This marsh has an area of several hundred acres.

The marsh in the valley of Stessing Pond contains a great body of peat: probably five hundred acres are underlaid by it, two yards deep; and its contents may be estimated at five hundred thousand cords.

On Mr. Hoag's farm, in Stanford, considerable peat is found. Professor Cassels reports that an extensive bed of peat occurs three miles east of Poughkeepsie. It is on the farm of Peter Van Voorhis, and contains about twenty-five acres. It is quite deep.

A small bog occurs near the red school-house, about two miles from Union Corners, in Dutchess County. Professor Cassels thinks the bog is deep.

In Clinton, four miles east of Union Corners, is a peat-bog of about sixty-five acres. It is on the land of Messrs. Underwood & Denison, and contains about sixty thousand cords.

A large deposit of peat was observed by Professor Cassels, two miles south of Union Corners, on the land of Elias Tompkins. It was estimated by him at ninety acres.

Mr. Merrick observed a small peat-bog, about two miles north of Hurd's Corner, in Pawling, containing probably five thousand cords.

Another, of nearly sixty acres, in the south-east part of Stanford, with a depth of about six feet.

Shaw Pond and Mud Pond, between Stanford and Washington, and Round Pond, in Washington, are filling up with peaty matter.

Rev. Mr. Shafter, of New York, observed peat in Rhinebeck, Northeast, and Clinton, in 1817.

It may be estimated that there are five square miles of peat in the salt marshes of Westchester, or three thousand two hundred acres, which will yield upon an average one thousand cords of peat of the second quality to the acre, or, in round numbers, three million two hundred thousand cords. The marshes of the Hudson River in New York, Westchester and Putnam Counties, that will yield peat, may be estimated at one thousand acres, with a yield of two thousand cords per acre, or two million cords. These include those near Sing

Sing, Verplanck, Peekskill, Anthony's Nose, Constitution Island, and numerous smaller ones. The peat in most of these marshes is supposed to be of inferior quality, fibrous, and containing much earthy matter. That formed in marshes in the interior of those counties is of much better quality, and far superior as an article of fuel.

On Joshua Raymond's farm, one fourth of a mile west of Bedford, in Westchester County, Professor Cassels reports that there is a peat-bog of sixty acres.

Also on Abraham Underhill's farm, two and a half miles south of Crumb Pond Village, Westchester County, Professor Cassels reports fifty acres, averaging twenty feet in depth.

A peat-bog of thirty or forty acres lies near the east side of Mahopack Pond, in Putnam County.

A large peat-bog was observed near Patterson, Putnam County.

A peat-bog of six or eight acres was observed in Phillipston, on the Phillips estate, about two miles east-north-east of West Point.

Another, on the same estate, about eight miles from Cold Springs, on the turnpike road to Putnam Court House, may contain fifteen or twenty acres.

Another, on the road from Putnam Court House, to Patterson, in Putnam County.

Another, four miles south-east of Peeksville, of six or eight acres.

Another, about one mile south of Scrub-Oak Plains, Westchester County.

Another, east of Stewart's iron mine, at the base of the hill.

Another, half a mile south of this mine, in the town of Phillipston.

Another, near Davenport's Corners, five miles north-east of Cold Springs.

Another, half a mile west of Saxon Smith's, in the south-south-east part of Phillipston.

There is said to be a fine deposit of about one hundred and fifty acres east of Croton, and about four miles south-east from Sumerstown Plains.

Several of the salt marshes in this region contain peat, mostly fibrous, and of inferior quality. That of finer texture and quality may probably be found at a depth of four or five feet. A moderate estimate of peat of second quality in these salt marshes would be five hundred thousand cords.

A peat-bog of about forty acres was examined by Professor Cassels, about one mile south of the Long Clove, in Rockland County. It is on land of Isaac B. Van Houten, and is supposed to have a mean depth of six feet.

Another is in the valley of the Hackensack River, about two miles west of Nyack, and contains about fifty acres, with a mean depth of six feet.

Peat, a few acres in extent, was observed by Professor Cassels at the north end of Rockland Lake.

Professor Cassels reports a peat-bog of forty acres on land of John Snediker, one mile south-west of Snediker's Landing, with an average depth of six feet of good peat. This peat-bog has been wrought for the New York market. It was opened during the summer of 1838, —

The proprietor receiving	25	cents per	chaldron.
The cartage to dock, ...	37½	"	"
Freight to New York, ...	37½	"	"
Expense of digging and curing, .	50	"	"
	$1.50		

It has been sold in New York at $3 per chaldron, or $4.50 per cord.

Professor Cassels observed a bog of fifty acres of peat between the lower village of Clarkston and the Hackensack River. Its average depth is stated to be six feet. It is on lands of William O. Blines, Levi J. Gurnee, and others.

Extensive peat-bogs were observed on the mountain, near the turnpike from Haverstraw to Munroe Works. One contains about forty acres, another about fifty acres.

Some other smaller bogs were seen, containing perhaps twenty thousand cords. Another was seen in the valley of Stony Brook, containing, probably, ten thousand cords.

At the north end of Long Pond, near the west line of Rockland County, a deposit of peat was observed.

A peat-bog of about ten acres occurs about one mile south-west of Stony Point.

Several small peat-bogs were observed near Fort Montgomery, that in the aggregate may contain ten thousand cords. There is a small one south of Fort Putnam, and near it; another south-east; another south-west; all within one fourth of a mile.

A peat-meadow occurs on the mountain, half a mile west of Round Pond, five miles south-west of West Point, on land of Mr. Wilkins, and contains ten thousand cords.

A small peat-bog was seen between the Limestone Ledge and Duck Cedar Pond, in Warwick, and may contain four thousand cords. Another was seen near the Patterson Mine.

Another near the Crossway Mine, and another east of the Starling Mine.

Another, of sixty to one hundred acres, in the valley of Smith's Clove, between Wike's and Galloway's.

A peat-bog lies west of Townsend's ore bed, in Canterbury, and contains ten thousand cords.

In the slate and graywacke region of Orange County, peat is everywhere abundant; and the localities are so numerous, that it would be tedious to enumerate them. The drowned lands, the Graycourt Meadows, and the Black Meadows, are the most extensive of these deposits. The former marsh is most extensive, and contains seventeen thousand acres. At a low estimate, there must be twenty-five thousand acres of peat-bogs in Orange and Rockland Counties that have not been estimated; and we may calculate, in round numbers, that they contain twenty-five million cords of peat.

In the drowned lands, several thousand acres are covered with this substance: it is from three feet to several yards in depth, and, on trial, proves to be a good fuel.

The Graycourt Meadows, lying in Goshen and Blooming Grove, contain five hundred acres of peat, several feet in depth. It exists in great abundance in Warwick, Minisink, Goshen, Monroe, Cornwall, Blooming Grove, New Winsor, Newburg, Montgomery, Hamptonburg, Crawford, Walkill, and Mount Hope; all the towns in the county, except Deerpark. In this latter town the quantity is small. The quantity in the county is unusually large in proportion to its extent; perhaps as much or more so than any other county in the State. It would require a great amount of time to ascertain the number of acres. It is perfectly inexhaustible. If its consumption for fuel ever becomes as general as it is in some parts of Europe, the Graycourt Meadows and drowned lands would prove a source of immediate and immense revenue to their proprietors.

Many smaller deposits of this substance are found in the towns of Schroon, Chester, Warrensburg, Johnsburg, Queensburg, Lake Pleasant, and Wells, varying in extent from one to five acres.

The summits of the Delaware and Hudson Canal, between Wurtzborough and Red Bridge, in Sullivan County, contain about fifty thousand cords.

A marsh extending down the valley of the Basher's Kill, from near Wurtzborough towards Cuddebackville, contains a thousand acres.

A bog a few miles west of Ellenville is said to contain about one hundred acres.

The bog in the valley of Three Brooks, south of Monticello, about four hundred acres.

The marsh one mile south-west of Monticello, belonging to Hon. Mr. Jones and others, about fifty acres.

The marsh half a mile south-west of Monticello, belonging to Hon. Mr. Jones, eight acres.

The marsh half a mile west of Monticello, belonging to Hon. Mr. Jones, say ten acres.

Several marshes between Monticello and Bridgeville, on the Neversink, perhaps fifty acres in all.

Bog between Dashville and Esopus, on the north end of the Passant Binnewater, say forty acres.

Bog one mile north of the above, on the north branch of the south fork of Black Creek, in Palts, Ulster County, perhaps ten acres.

Bogs on Black Creek, near the Poughkeepsie and Palts Turnpike, about fifty acres.

Bog in the valley of the east branch of the Delaware, in Roxbury, Delaware County, on Mr. Straton's farm, about ten acres.

Bog in Marlborough, on the south road from Marlborough to Pleasant Valley, on land of G. Birdsal, Mrs. Bingham, and D. Cassman, in Ulster County, about twenty acres.

A few deposits of this substance have been discovered in Otsego County, on the farm of Mr. Clark. Two miles south of Cooperstown is a small peat-swamp, nearly surrounded by alluvial hills, covering an area of six acres. The peat is apparently of good quality, and may be made accessible with little expense. Several other small marshes of a similar character are found in this neighborhood. On the road from Cooperstown, west towards Oakville, are several small deposits. Two miles south of New Berlin, on the east side of Madilla, is a peat-swamp, covering an area of forty-five or fifty acres. This is the most extensive deposit of the kind of which we have any knowledge in this region. Its depth is unknown: in several places a pole was thrust down ten feet. It is probably much deeper.

In the towns of Great Valley and Little Valley, the "sags," or depressions in which the clay is formed, contain more or less extensive bodies of peat. The largest is upon the land of Mr. Sweetland. About ten acres are spread over by the bog; and the depth of peat varies from a foot or two near the margin to more than twelve towards the centre.

About ten acres between Reed's Mineral Spring and Argyle, in Washington County; fifty acres at the south end of Summit Lake, in South Argyle, and one or two other deposits, comprising perhaps thirty acres, in the same town; twenty-five acres on A. McNeil's farm, in N. Greenwich; ten acres on P. Reynolds's farm in the same town. About one hundred acres lie between N. Argyle and Hartford; seventy acres one mile south-east from N. Greenwich; ten acres in Greenwich, about a mile and a half south-west from Lakeville; and probably about four hundred or five hundred acres in various places in Greenwich, about a mile south from Lakeville;

ten acres on Stephen Durham's farm; and about ten acres on Simeon Burton's land, in Easton. In Salem, about three miles north from the village, is a tract of about one hundred and sixty acres; and about four miles north-east from the village is a tract of one hundred and thirty acres. In West Hebron we find about twenty acres, two miles south-west from the village; and in Kingsbury, on the Champlain Canal, lie something like a thousand acres. At Fort Ann, two miles north-east from Mount Hope Furnace, is about thirty acres; and between the two lakes south of Mount Hope lie about fifty acres. In Dresden are several deposits, amounting to about one hundred acres. In Putnam, four miles north-west from the ferry, there are about thirty acres. In Greenbush, on the Boston Turnpike, is a small deposit; and another of a few acres on land of Mr. Clark, at Coyeman's, three miles from New Baltimore, in Albany County. In Hamilton County, Marion River, about seven miles in length, connects the Eckford Lakes with Racket Lake: it passes through an immense marsh containing an inexhaustible quantity of peat, and upon which, as in numerous other localities, there is a great amount of tamarack, especially valuable for ship-timber.

Bones of the mastodon were found in 1790 and 1800 in the town of Montgomery, about twelve miles from Newburg, in Orange County, ten feet below the surface, in a *peat-bog* in marl. Several bones of the legs, some of the vertebræ, several ribs, and some of the bones of the head, were obtained. One of the leg-bones measures more than forty inches round the joint, and thirty-six on the cylindrical part of the bone, and is nearly five feet long. The teeth are nearly seven inches long, and four broad: they are found white, and fast in the jaw, without appearance of decay. The holes in the skull, where the nostrils appear to have been, measure eight inches in diameter.

Eight similar skeletons have been discovered within eight or ten miles of this point, and some of them were fifteen or twenty feet below the surface of the earth.

Some bones of these animals were found in 1782, three miles south of Ward's Bridge, in Montgomery; another, one mile east of this bridge; another, three miles east; another, seven miles north-east; another, seven miles east; another, five miles westwardly; and another, ten miles north, in Shawangunk. Similar remains were found in digging the Delaware and Hudson Canal, in a *peat-bog* between Red Bridge and Wurtzborough, in Sullivan County. Fossil bones of the mastodon and elephant have been found in other parts of Orange County.

These facts are gathered from a paper by Sylvanus Miller, published in the "Medical Repository," New York, 1801. Mr. Miller in a letter to De Witt Clinton in October, 1814, narrated some of them, and mentioned the additional circumstance, that, in some cases, considerable locks and tufts of hair were found with the skeletons. It was of a dunnish-brown color, from one and a half to seven inches in length, and retained its natural appearance until exposed for some time to the air, when it gradually mouldered away into a kind of impalpable dust.

Mention is made of the skeleton of a mastodon found in Genesee, in marl, below a bed of *peat*.

In September, 1866, the jaw-bones of a mastodon were discovered in a *peat-bed*, some thirty feet below the surface, at Cohoes. The length of each jaw-bone is thirty-two inches; the breadth across the jaw, at the broadest point, twenty inches, and the extreme depth about twelve inches. On one side is a tooth four inches in length and two and a half in width, and on the other side two teeth, one of which is six and a half inches long, the other four, and both uniform in width and shape with the tooth opposite.

On the side of the single tooth there is no cavity to indicate that any other ever existed, and it is evident that the animal died in possession of all the dental conveniences that nature ever gave him. The front of the jaw is comparatively light, showing that the creature was not given to carnivorous diet, but subsisted chiefly on herbs.

About a month later, most of the remaining bones of the skeleton were discovered and exhumed. They lay at a depth of eighty-five feet below the surface, and consisted of the back-bone, a number of the ribs, the hip-bones, shoulder-blades, and bones of the hind legs, two tusks, the upper jaw, and cranium.

The tusks were each nearly six feet long, and about nine inches in diameter. One of them, upon exposure to the air, crumbled to pieces like clay, resembling that substance in appearance and texture. The ribs, of which fourteen were found, are about four feet long, the largest being four feet nine inches. The upper jaw-bone is four feet nine inches long from the extremity of the mouth to the cranium; across the forehead measures about three feet. So heavy is the skull that it was with difficulty that four laborers could move it. The sockets in which originally were located the eyes of the monster, are almost large enough to admit the head of a man. The hip-bone is five feet long, and weighs one hundred pounds; the shoulder-blades measure two feet nine inches, and weigh about fifty pounds each; the bone of the leg, at the knee-joint, measures thirteen inches in diameter; the vertebræ of the back-bone are eight inches in diameter. The other fragments found are in harmonious proportion to those already mentioned.

The bones have been carefully prepared, the missing parts ingeniously replaced with wood, and the whole set up and united by means of wires, and placed in the State Geological Museum at Albany.

We understand that in the same *peat-bed* was likewise discovered an ostrich egg, at about thirty feet below the surface.

The peat was covered with a slate rock, which was cut through in digging foundations for a mill. On each side of the peat-bed are perpendicular rocks, in which are large semicircular cavities, evidently worn by the action of water, and showing that a stream once flowed over the spot.

On Long Island, throughout its entire length, there are numerous peat-beds, some of them of great extent, most of them of rather unusual depth, if we are rightly informed, say from ten to forty feet; and nearly all of them are believed to be of quality adapted to make, when properly manufactured, good fuel.

Within the last two years, extensive explorations, and, in some cases, very large purchases of peat-lands have been made, some for actual use and development, and others unquestionably for speculative purposes; and so clearly manifest is it that stores of solid wealth are to be found in these hitherto waste-places, that several of the leading railroad and manufacturing interests, as well as numerous private parties, are committed to the development of this fuel question, and are preparing for extensive operations the present year.

New York, in the length and breadth of the State,

may be said to be wide awake on the subject of peat-fuel. Large quantities will be manufactured the present year; and the day cannot be far distant when her peat-beds will be added, as no inconsiderable item, to the list of her already abundant resources of wealth.

New Jersey.

The deposits of peat in this State are numerous and rich, and a lively interest is manifested in utilizing them. It is well known that numerous shell marl-beds or pits are exclusively and profitably worked for fertilizing purposes. These marl-beds are usually found to rest upon a bed of clay, sand, or gravel, and are succeeded by peat or muck, the depth of the peaty deposits varying, according to the statements we have seen, from three to twenty feet. Peat, varying in quality for purposes of fuel, is found in Montague, Harrisville, Belleville, Hohokus, Sandiston, Vernon, and other localities.

It will be interesting in this connection to note something of the cedar-swamps, which are a remarkable feature of Southern New Jersey. They are common in all the counties south of Monmouth; but probably the most extensive are in Cape May, and the adjoining parts of Cumberland and Atlantic Counties. The Cedar-Swamp Creek, which runs into Tuckahoe River, and Dennis Creek, which runs into Delaware Bay, head in the same swamp; and the whole length of the two streams — a distance of seventeen miles — is one continuous cedar-swamp. The wood is the white cedar. It grows on peat, and its roots run near the surface. In the present growth of standing timber, scarcely any trees are to be found more than one hundred years old; but these rest upon a formation containing we know not how many generations of trees which have lived and fallen before them. Large stumps are frequently found standing directly on other large logs, and with their roots growing all around them, and then other logs still under these; so that one soon becomes perplexed in trying to count back to the time when the lower ones were growing.

Dr. Beesley, of Dennisville, some years since, communicated to the newspapers an article on the age of these cedar-swamps, which was copied by Professor Charles Lyell, in his "Travels in the United States;" in which Dr. Beesley says that he "counted 1080 rings of annual growth between the centre and outside of a large stump, six feet in diameter; and under it lay a prostrate tree, which had fallen and been buried before the tree, to which the stump belonged, first sprouted. This lower trunk was five hundred years old: so that upward of fifteen centuries were thus determined, beyond the shadow of a doubt, as the age of one small portion of a bog, the depth of which is as yet unknown."

Mr. Charles Ludlam counted seven hundred rings of annual growth in an old tree, which was living when cut down. The trees stand very thick upon the ground, and grow rapidly at first; but as they increase in size, and crowd each other, the tops become thin, and the annual growth exceedingly small. The rings near the centre of a large cedar log are often almost an eighth of an inch in thickness, while those near the bark are not thicker than paper. The soil in which these trees grow is a rich black peat, composed of vegetable matter, which, when dry, will burn, leaving not more than three or four per cent. of an ash almost insoluble in acids, and without the slightest alkaline taste. This peat is of various depths, from two or three up to twenty feet or more; and the trees which grow on it have their roots extending through it in every direction near the surface, but not penetrating to the solid ground. When this soil is opened to the sun and rains, it decays rapidly; but when covered with a growth of trees, and so shaded that the sun does not penetrate to the ground, it increases rapidly, from the annual fall of leaves, and from the twigs and small trees which die and fall. Trees are found buried in this peaty earth, at all depths, quite down to solid ground. The buried logs are generally quite sound: the bark on the under side of many of them is still fresh in appearance. The color of the wood is preserved, and its buoyancy retained. When they are raised, and floated in water, it is observed that the side which was down in the swamp is uppermost. These logs are so abundant in some parts of the swamp, and in the salt marshes bordering on them, that a large number of men are constantly occupied in raising and splitting them into rails and shingles. Our authority states that in Mr. Ludlam's swamp this business was commenced fifty years ago, and has been carried on every year since; the annual product sent from Dennisville alone, averaging, for a number of years, not far from 200,000 rails, worth $8 to $10 per hundred, and 600,000 shingles, worth $13 to $15 per thousand. The size of the logs raised is generally from one and a half to three feet in diameter, though four feet is not uncommon; and they have been taken out five, six, and even seven feet in diameter. In searching for logs, the workman uses an iron rod, which he thrusts into the mud until it strikes one; then, by repeated trials, he judges of its direction, size, and length: the stumps, roots, and earth are removed from over it; it is loosened by means of levers, and soon rises, and floats in the water which has accumulated in the excavation.

PENNSYLVANIA.

Of peat in Pennsylvania, the great coal-bearing section of this country, we shall hardly be expected to say much; but some remarks contained in the Geological Report of that State, published in 1858, upon the analogies of origin and structure of peat and coal, are of so much interest, that we shall make brief extracts.

"Peat-bogs are nothing but beds of coal, not entirely ripe or burned out. We have to mention some very interesting and striking analogies, such as the presence of bitumen in both formations. It has been asserted that bitumen could not proceed from plants, but it is everywhere obtained by the distillation of peat.

"We would show the identity of the geographical distribution of both formations, though it has been long asserted that the peat formation belongs to a cold climate, and the coal to a warm one; also the affinity of forms in their plants,—forms particularly adapted, it seems, to the absorption of the vapors, and the transformation of carbonic acid in woody matter; and show the large amount of wood in the plants of our peat-bogs, even in the mosses (*sphagna*), which, though very fragile, soft, and thread-like plants, have a larger proportion of woody matter than the hardest of trees.

"The analogy of formation being established, we do not see any reason to theorize about the general flora of the period of the coal formation.

"Our peat-bogs have more than seventy species of mosses, five or six species of *lycopodiaceæ*, and as many species of ferns, eighteen or twenty species of palm-trees, reeds, and phanerogamous monocotyledonous plants; and our peat-bogs are now essentially a compound of such plants, sedges, grasses, reeds, &c.

The trees that are now living on these bogs — viz., the pines, birches, &c. — probably take the places of the large series of palm and fern trees of the marshes of the Old World."

After making the above quotation from a French author, the report continues : —

"The above was written ten years ago : since then, the palæontology of the plants has made very great progress, and a great many new species have been discovered. Nevertheless, this analogy in the vegetation of both formations — the coal and the peat — has become more and more evident.

"In the peat-bogs of North America, ordinarily named cedar-swamps, there are, first, about twenty species of mosses of which the growth directly contributes to the formation of the peat." They are enumerated.

"Of these species, one only is peculiar as an American plant; all the others, without exception, have the same plan, the same destination, and are found in the same abundance, as in the peat-bogs of Europe. There are, second, many other species of mosses which are growing only on the peat, but which do not claim a large share in the combustible matter.

"The sedges and grasses are distributed, also, in the same proportion.

"The trees have some species which cannot be compared; but a great many of them, the birches and alders, are nearly alike.

"The type of the flora of the peat-bogs is everywhere the same, and is susceptible of proof, as well from our explorations in the great bogs of Southern Virginia as from the observations of travellers in the Australian hemisphere.

"It has been long asserted that the peat-bog formation belongs particularly to cold climates, and that the preservation of the woody matter is essentially due to low temperature. Our researches in Europe on this subject have already proved that this is not the case, and that the area of the peat-bogs occupies exactly the same latitudes as that of the oldest coal formations. And since we have been enabled to pursue the same explorations in America, we have found on this continent exactly the same distribution; for in this country the peat-bogs are found from far above the northern shores of Lake Superior, as high as 60° of latitude, to the Great Dismal Swamp in South Virginia, 35°, exactly in the same latitudes as are occupied by the great coal basins of America. We have made on the same subject very long researches.

"But how is it possible to account for the vegetation, in our latitude, of those immense trunks of trees, perhaps of fern-trees, to which we find an affinity only in the tropical regions, if we do not admit of a great change of temperature?

"In the peat-bogs of northern countries, — of Denmark, Sweden, and Switzerland, — we find sometimes, heaped in very thick strata, much larger trunks of trees than those which have been found in the coal.

"The following is a description, given in 1846, of a peat-bog which we visited, near Waldsmarslund, thirty miles above Copenhagen: 'These deposits of wood are a true forest heaped upon another, and buried in the peat, which in the marshes of Denmark are found everywhere with alternating beds of the same species of trees. At the bottom, upon a bed of peat from four to six feet thick, are *pine-trees*, lying flat, nearly always in the same direction as the inclination of the basin; viz., the roots against the borders. These pine-trees are ordinarily from six to ten inches in diameter: their slightest branches are preserved, and they are embedded in a mass of their own leaves, cones, mushrooms, &c., of which the form is not at all altered. Upon these pines is a bed of black peat, from five to six feet thick, overlaid by a forest of prostrate white birches. Upon the birches there is again six feet of less decomposed peat, covered with enormous trunks of oaks, which have no less than three feet of diameter, and of which the wood is so well preserved that it is sawed on the place, and used for building material. The peat in which these trees are buried does not preserve any trace of the leaves of these oaks, but only some acorns. It is evident, nevertheless, that they have grown on the place where they are found, being preserved in their integrity, with their smallest branches and their bark. These trees are covered by from six to eight feet of peat, in which or upon which is found sometimes a fourth forest of trees, and this time of beech-trees, the same trees that now form the forest around.'

"The size of the trunks buried in the peat-bogs is, as is easily seen from the above description, in favor of our present formations. In the Dismal Swamp of Virginia, we have seen, in the peat, trunks of magnolia measuring more than one hundred feet, without great diminution in their diameter. The trees of the coal which are ordinarily ascribed to a genus of plants like the ferns, and so to fern-trees, were not true ferns, but a peculiar species of plants, of which we have no living representatives, and of which the nearest relatives are the lycopodiaceæ, a genus of plants of which the largest species known are living in the peat-bogs and the forests of our northern hemisphere.

"Our human race is young, but the world lived long before it. Truly, Nature has prepared our home. It has heaped, for the future welfare of our race, those inexhaustible beds of combustibles that afford us so much comfort; but it has done this without any miracle, without any of those sudden transitions which we are so prone to discover for the satisfaction of our own pride. This coal formation" (and the same is equally true of peat) "is an admirable one, and we can look to it only with wonder, and with faith in an overruling Director; for this heaping of combustible matter was by itself nothing but a useless proceeding. The bed to put it in was to be prepared, a long time in advance, by a thick layer of clay, to prevent the egress and the dispersion of the bitumen after its separation from the woody matter. It required, also, an impermeable covering to prevent too strong an action of the oxygen, and the mingling of sand and other strange matters, which would have entirely changed its combustible property. This, as well as the formation of iron, which is also ordinarily in progress with the formation of the peat-bogs, has been obtained by the simplest laws and the slow progress of Nature.

"This proceeding does not come to its end without a great many changes.

"We have seen some beautiful illustrations of these changes in Germany and Denmark. Near Leipsic there is a bed of lignite, formed of large trunks heaped about fifteen feet thick. The matter is entirely soft, and all the trunks flattened, measuring in one direction scarcely half the diameter they have crosswise. It is also entirely black, and yields an excellent fuel. It is extracted with shovels, like peat, after its surface has been laid bare from twenty feet of sand and gravel lying upon it.

"In Denmark, about twenty miles below Copenhagen, near the sea-shore, there is an extensive plain, covered with the finest grass, and affording excellent pasturage to large herds of cattle. By digging there, they find, under one foot of humus, a bed of *peat*, entirely composed of bark of birches. This bark is heaped six feet thick, and closely packed and flattened. It is cut out and dried in long rolls, entirely void of earthy matter. The woody matter, nearly fluid, or transformed into a very soft yellow mass, is at the bottom of these beds, and is taken out of the excavations with buckets, spread on layers of straw, through which the water percolates, and, when thickened, it is beaten hard, dried, and burnt like coal."

VIRGINIA.

The peat interest of this State may be said to centre in the Great Dismal Swamp, which, it is ascertained, contains some of the most extensive and superior deposits of peat of which we have any knowledge in this country.

The Virginia Condensed Peat Company and the Great Dismal Swamp Peat Company have already taken the initiatory steps, and we hear of other companies and individuals who are operating with a view to the development of this immense fuel region. The last-named company, in their published statement, say, —

"The peat of the Dismal Swamp has been the growth of uncounted ages. Recent geological investigations have established the fact, that 'the Bare Garden,' which is the richest and best portion of this enormous peat-field, was once covered by a gigantic forest of resinous woods, principally the gum-tree, cypress, juniper, and pitch-pine, which flourished in primeval luxuriance. When these forests were prostrated by convulsions, or the slow process of decay, they were decomposed and covered by mosses and grasses, which accumulated for many centuries, until, by gradual chemical changes, the whole became one vast homogeneous mass of perfect peat, extending to undiscovered depths, and richer in calorific ingredients than almost any other peat hitherto discovered."

Samples of these peats were furnished to Professors Johnson and Silliman, of Yale College, for examination and analysis, and their reports upon them are as follows: —

"SHEFFIELD SCIENTIFIC SCHOOL OF YALE COLLEGE,
"NEW HAVEN, CONN., November 27, 1866."

"GENTLEMEN:

"I have examined the two samples of peat sent by you as coming from the Great Dismal Swamp. These samples, brought to perfect dryness, gave, —

	Surface Peat.	At 4 ft. depth.
Volatile matter	62.56	65.74
Coke (ash-free)	31.21	31.81
Ash	6.23	2.45
	100.00	100.00

"Calculated with the usual content of water (twenty per cent.) occurring in air-dry peat-fuel, we have, —

	Surface Peat.	At 4 ft. depth.
Water	20.00	20.00
Volatile matter	50.05	52.59
Coke (ash-free)	24.97	25.44
Ash	4.98	1.97
	100.00	100.00

"Both of these samples are of excellent quality. The sample taken from a depth of four feet, especially, is remarkably free from ash, which adapts it for gas-making, metallurgical purposes, &c. For iron smelting and working, this peat, properly condensed, would answer admirably.

"Very truly yours, &c.,

"SAMUEL W. JOHNSON,

"Professor of Agricultural and Analytical Chemistry, Yale College."

"GENTLEMEN:

"My examination of the specimen of peat from the Dismal Swamp, which you placed in my hands, has yielded the following results: —

"As received, the peat appears like a thick mud, dark brown in color, impalpable in fineness, crossed here and there with root fibrils, and remarkably uniform.

"Annexed are the results of its analysis: —

	I.	II.	III.	IV.
Water . . .	78.89	20.22	——	10.00
Volatile matter .	13.84	52.31	65.53	58.08
Charcoal . .	6.49	24.52	30.82	28.59
Ash	.78	2.95	3.65	3.33
	100.00	100.00	100.00	100.00

"No. I. is the wet peat, as received; II., the same, after drying for six months in my laboratory; III., the perfectly dry peat; and IV., another portion, calculated with ten per cent. of water.

"From these figures it is at once evident that this peat has an excellent composition. The large amount of combustible matter it contains, and the correspondingly small proportion of ash, greatly enhance its value as a fuel. Moreover, the percentage of coke which it yields is so considerable as to prove that its conversion into charcoal for use in the manufacture of iron, for example, may be made remunerative; and especially so for the above purpose, since an analysis of the ash failed to show the presence of any injurious substances.

"I regard the physical and chemical examination to which I have subjected your peat as highly satisfactory. And I do not doubt that, by the use of the approved modes of preparation which you are employing, you will be able to place in the market an exceedingly useful and merchantable fuel.

"Yours respectfully,

"B. SILLIMAN.

"YALE COLLEGE LABORATORY,
New Haven, Jan. 1, 1867."

We have ourselves had samples of this same peat, and applied to them the practical tests which every purchaser of fuel applies to wood or coal, in order to satisfy himself of its real value, or, in other words, how much money he is willing to pay for it.

The scientific analyses above reported confirm our practical tests, and we think there can be no doubt that the extensive deposits of peat in the Great Dismal Swamp are among the purest and best in this country.

An officer who served during the late war, and was for a considerable time stationed in this region, and improved his opportunities for observation, writes, —

"There can be little doubt that the Great Dismal Swamp was once a vast fresh-water lake, below the level of the neighboring lands, fed by the springs and rills therefrom. Gradually the mosses and dense sedges vigorously growing in the rich soil of its quiet margin encroached on all sides, increasing in height and firmness, — fed by drenching rains and frequent fogs, until they have nearly filled the huge basin, — and crowding the waters inward and upward to an elevation of twelve feet above the level of the lands at the verge of the swamp, hold Lake Drummond compressed to one tenth its original size, and imprisoned in forest and bog.

"The broad canal built by the U. S. government through the eastern part of the swamp is fed by a raceway from the lake; and the lake is so much higher than the canal, and the middle of the canal so much higher than the ends at the edge of the swamp, that the water flows from the lake into the canal, and each way into the Elizabeth and Pasquotank Rivers.

"The graphic delineation of Porte Crayon, and the startling romance of Mrs. Stowe, alike fail to convey an adequate conception of the weird wildness of this unique locality. Equally impossible is it to estimate and describe the exhaustless wealth of this mysterious spot, in the choicest timber and in the richest and purest of American peat.

"In America the dryness of the air, the rareness of fogs, and the heat of summer, render the climate less favorable than that of Europe for the normal production of peat by the steady growth of vegetation.

"On the cool and humid coast of Labrador, in foggy Nova Scotia and Newfoundland, and in the Great Dismal Swamp of South Virginia, with abundant moisture from its lake, and the dense vapors from the adjacent ocean, peat formations bear close resemblance to the huge bogs of Ireland and Germany.

"Nature seems to have provided fuel where it is most wanted. Below 35° of latitude little fuel is needed, and wood grows with great rapidity; above 60° few live to need fuel. Whether or not this be the quiet reasoning of busy Nature, or her divine Author, certain it is that above the limit named the cold interferes with the requisite growth and decomposition of the proper vegetation, while below that limit the heat and dryness of the air largely absorb moisture, and cause a decomposition too rapid for the formation of peat. The Great Dismal Swamp is substantially the southern limit of peat in this country. It may well be doubted whether peat of the finest quality would have been formed there had not the vast size of the swamp, the abundant water in its large lake, and the added moisture of the dense and frequent fogs from the adjacent ocean more than neutralized the effect which the heat of the climate and the dryness of the air would have produced upon a smaller inland swamp.

"The water of the Great Dismal Swamp, though of a dark reddish color, is clear and fit for all domestic purposes. I have often used it to quench my thirst, in preference to the water of the springs or wells of that vicinity. Indeed, the people near the swamp believe it to possess medicinal qualities, and declare that those who steadily use it will not be troubled with ague or the bilious disorders incident to summer in that region.

"I have heard the inhabitants of the neighborhood repeatedly affirm that to visit the swamp and partake of the waters beneath the dense shade of its luxuriant magnolias, lofty junipers, and white cedars, was as invigorating as a trip to the Sulphur Springs or the healing waters of Saratoga. We smile, as if the idea was absurd; but certain it is, that the juniper water, as it is called, possesses as stringent qualities, at least, — and the experience of the army abundantly proves that it is much more wholesome to those not acclimated, — than the waters of various villanous properties, drawn from the wells of that torpid locality. To us the huge, strange swamp itself was much less dismal and unpleasant than the flat, half-tilled border land of our encampment."

From a letter received from the superintendent of the Virginia Condensed Peat Company, we make the following extracts: —

"The juniper, cypress, gum, poplar, and other valuable woods, so common in this swamp, have long been a source of trade; from which cause parts of the swamp are wholly or nearly cleared of its heavy growth. Extensive fires, which have occurred more or less every year, have served also to clear the way for peat operations.

"The peat is of the first quality, as proved by careful analysis. It varies in depth from four feet to forty or fifty feet. In the locality occupied by our company it is *ten* feet deep and upwards.

"The facilities for getting the peat to market are good. The Dismal Swamp Canal, which runs through, not exactly the centre, but the eastern part of it, affords a water communication. Vessels of light draught may come alongside the works, located on the banks, take in the peat, and sail for any port they please, north or south.

"Many get the idea from the name of the swamp, "dismal," that it is a gloomy and unhealthy place. It is not so. It is more pleasant and beautiful to the eye than the low, sandy, level plains that border upon it, and *far more* healthy. Negroes have been known to live for years in the swamp, enjoying excellent health and strength. 'Lumbermen,' says Mrs. Stowe, in her thrilling romance of the Great Dismal Swamp, 'who spend great portions of the year in it, cutting shingles and staves, testify to the general salubrity of the air and water. The opinion prevails among them that the quantity of pine and other resinous trees that grow there impart a balsamic property to the water, and impregnate the air with a healthy resinous fragrance, which causes it to be an exception to the usual rule of the unhealthiness of swampy land.'

"It is even so. Such is the universal belief of the inhabitants, both black and white, who live in the vicinity of the swamp. The water is clear as crystal, though of a reddish color, caused by the abundant growth of the tree which gives it its name, 'Juniper water;' it is sweet and healthy, and used for all culinary purposes. In no part of this vast swamp is stagnant water to be found. Like the water in all perfect peat-bogs, it is pure and free from malaria.

"In digging the peat it might naturally be supposed, owing to the woody growth of the swamp, that the roots would be a great hinderance. These roots are near the surface, and are no very serious detriment. At the depth of two or three feet, you are free from them; then the peat comes pure, of the consistency of butter or lard, and unlike most of the more northern peat, it contains no fibrous roots. From our limited experience thus far, we perceive no material difference in the quality of the peat, whether lying near the surface or at a greater depth.

"Owing to the situation of the swamp, at the southern limit of the peat region, the season for manufacturing it comprises *nearly the whole year.*"

Another correspondent, an officer, who likewise served in this region for several months, and was so much impressed with the value and importance of it, that he has since revisited it, and devoted a good deal of time to explorations and investigations, writes as follows: —

"With the exception of a single locality in Virginia, I have been unable to learn of any deposits in the Southern States.

"Decomposition of vegetable matter is so rapid in all warm climates as to prevent the formation of peat. In this case, however, a combination of natural causes produces a climate similar in most respects to that of Ireland, and consequently peat is formed there equal in quality to the finest varieties of the Irish bogs.

"This deposit is situated a few miles from Norfolk, on the line of the Dismal Swamp Canal, in the vicinity of Lake Drummond.

"Heretofore these swamp lands have been valued only for their timber; and for at least a century the cypress and juniper logs, dug from beneath the 'black bog' (as the negroes call it), have constituted quite an important article of commerce.

"It is only within two years that the discovery has been made that the peculiar preservative properties of this 'bog' were identical with those of the Irish swamps, and that this vast waste of saturated, spongy substance contained all the elements of a most valuable fuel. So soon as it became known, gentlemen from New York and Baltimore secured all the available tracts, and are now preparing to work them. These tracts comprise in the aggregate about fifteen thousand acres.

"As the depth of this deposit is in no case less than six feet, and in many places from forty to sixty feet, it will be readily seen that Nature has established here an immense depot for the supply of fuel to all the seaboard cities. As the entire transportation will be by water, the expenses of marketing the fuel will be materially less than though land carriage were necessary.

"Samples of this peat in a crude state have been analyzed by distinguished chemists, and I understand that you have their reports.

"From them, it will be seen that the Dismal Swamp peat is not only a valuable fuel, but that from its unusual purity it is especially adapted to the manufacture of illuminating gas, and for all the purposes of metallurgy.

"It would not be surprising, should the capitalists now engaged in developing the rich deposits of gold and copper in North Carolina and Virginia find that the peat of the Great Dismal Swamp was the best and most economical fuel they could employ for the smelting and desulphurizing of their ores.

"One peculiarity of this swamp is its entire freedom from malaria. While on the open lands surrounding it, ague and congestive fevers prevail, none has ever been known within it, although there are lumbermen now at work who have spent fifty years among its recesses. During the prevalence of yellow fever in Norfolk, the swamp proved a safe retreat for hundreds of people.

"There is no doubt but that this salubrity of climate is due in a great degree to the balsamic nature of the principal trees with which this vast area was once entirely covered.

"The water of the lake and swamp is of the color of sherry, and, although perfectly quiescent, it never stagnates, but remains perfectly pure and sweet.

"For a great many years the naval authorities at Gosport have been in the practice of filling the water-tanks of all vessels bound on long voyages with this juniper water, and sailors have never known it to spoil."

The concurrent testimony, from all these sources, in regard to the "sweet water" of this swamp, is of interest as bearing upon the question of the antiseptic properties of peat, which will be more fully referred to in succeeding pages.

We have credible information of a barrel of this water which has been in the possession of a gentleman for more than thirty years, which is apparently as sweet and clear now as when first taken from the swamp.

Ohio.

We have, at present, but very little information relating to the deposits of peat in this State; have learned of but few, and those mostly on the lake shore. We are seeking information from this section, aud hope to be able to present, in a subsequent edition, some detailed statements.

Michigan.

In a letter recently received from Professor A. Winchell, of the University of Michigan, at Ann Arbor, he says, —

"I have not been unobservant of your efforts for a few years back to introduce machinery that should enable us to utilize the vast deposits of peat existing in various parts of our country. I have anticipated that the time would arrive when the existing deposits of the lower peninsula of Michigan would come into requisition.

"I am induced to think that the time has now arrived. Cord-wood commands, throughout the southern portion of the State, not less than five to eight dollars per cord; and the coal which is worked at Jackson and Corruna and other points, though obtainable for three to four dollars per ton, is not of quality suitable for domestic use. In this state of things, we are introducing coal from Pennsylvania, — Scranton, Lackawana, and Lehigh, — at twelve to thirteen dollars per ton delivered. This coal is cheaper than cord-wood; but its use involves an unreasonable expense, as long as we have, within the limits of every county of the State, a deposit of combustible matter sufficient in extent to supply the wants of the people for a long time, provided we have the means for effecting its inexpensive desiccation and preparation for use.

"The peat-beds of the State were described in a general way in the Geological Reports made by Dr. Houghton and his assistants, thirty years ago.

"Eleven years ago I published an account of them in the Michigan Farmer, then edited in Detroit. In 1860, I made further mention of them in my Reports on the progress of the Geological Survey; and in January, 1865, in an address before the Legislature, at Lansing (also published by the Legislature), on the 'Soils and subsoils of Michigan,' I again made reference to the subject.

"In none of these documents, however, are our peat accumulations described in detail. The notices embraced in the Geological Reports are general and hasty, such as would be suited to reports of progress only. No *final* Report has ever been made, and hence no opportunity offered to record the numerous details of which observations have been taken.

"The facts are, however, that few States in the Union afford so extensive accumulations, or so fine a quality of combustible muck and peat.

"I cannot now specify many of these accumulations of peat, neither is it necessary. I know of very extensive beds in Washtenaw, Jackson, and Lenawee Counties. I have not made special explorations of many of these beds. There has been no occasion for it.

"Whenever the time arrives for drawing upon these stores of fuel, local observations can be made, and measurements of depth and tests of quality undertaken.

"You seem to incline to the opinion that the peat deposits in Michigan are rather thin, and not of very good quality. In this you are quite mistaken. The fact is, as I have already stated, our peat deposits are numerous, often from four to twenty feet deep, and of good quality.

"I know of some peat-bogs of forty to one hundred acres in extent. In Ingham County is one of more than a thousand acres in extent; another in Lapeer County. As nearly as I can recollect, a bog occurs in Washtenaw County, one or two hundred acres in extent; another in the south part of Jackson County, fifty acres in extent.

"I am unable to say what depth the peat attains in these particular places, but I know that in many other places it is ten or fifteen feet to the bottom."

The Detroit Free Press, in a recent article, says, —

"The discovery of immense beds of peat in Michigan, which are already being developed, and the fact that the article is now produced for commerce, is a matter of great importance to every housekeeper, as well as to the capitalist and manufacturer.

"Specimens of crude and condensed peat have been forwarded to the Board of Trade in this city, for exhibition at their rooms, by Mr. Elisha Congdon, an enterprising merchant and farmer of Chelsea, in this State. The bed from which the specimens were taken is situated in Chelsea, Washtenaw County. It is what is known as a dry bog, containing about one huudred acres, with an average depth of ten or twelve feet at least. Soundings have been made of some parts of it, giving over twenty feet, without reaching the bottom of the bed.

"Professor Douglas, of the Michigan State University, after a thorough analysis, pronounced it a very superior article, yielding only three and two tenths per cent. of ash.

"These beds of peat are almost inexhaustible, and being found in the dry bogs, can readily be worked; and the quality of the material having been fully tested, we may expect soon to have our markets supplied with it as a regular article of traffic.

"Mr. Congdon is now erecting a building in which he will place one of the Leavitt peat machines, and expects to prepare for market, during the coming spring, fifteen to twenty tons per day of the concentrated peat.

"To what extent this traffic may reach is beyond calculation; for it is believed that throughout the State there are hundreds of peat-beds which a proper system of drainage will develop, and thus furnish an abundant supply of fuel, when the productions of our forests will have become curtailed to such an extent, and its cost as fuel so enhanced, as to prohibit its use for that purpose.

"We hope that the owners of the numerous peat-marshes in the county will profit by the example of Mr. Congdon."

INDIANA.

It is well known that there exist, in various parts of the State, immense tracts of swamp-lands; and that during the past few years, large portions of these have, by a systematic method of draining, been reclaimed for cultivation, and have proved very valuable for that purpose.

Some thousands of acres of these lands, it is now reported, are found to be deposits of peat, of good quality for fuel, and operations on a limited scale for manufacturing it have been commenced at one or two points during the past season.

We have seen it stated that along both sides of the Kankakee River, extending from South Bend to the Illinois line, is a peat-bog of more than sixty miles in length, with an average width of three miles. In some places it is known to be over forty feet deep; but even though it average only one half or even one quarter of this depth, the aggregate amount of fuel it contains, it must be admitted, is beyond comprehension.

It is, we understand, quite practicable to drain this entire marsh to the depth of ten or fifteen feet, at small expense.

South of the Kankakee, the peat-bogs between there and the Wabash are simply immense, and they are traversed by three railroads.

Near the head of Lake Michigan, within ten to fifty miles of Chicago, there is peat-fuel enough to supply that city, if it lives to the age of Jerusalem. The marsh along the Calamie River will measure at least thirty miles, averaging a mile wide, and is of unknown depth, and undoubtedly good peat.

There would seem to be no good reason why these rich and extensive deposits should not be improved for fuel, to the pecuniary profit of individuals and greatly to the advantage of "the people."

ILLINOIS.

In the northerly part of this State are valuable deposits, some of them quite extensive; but we are not, at present, in possession of sufficient reliable data to make mention of them in detail.

From a lengthy and somewhat elaborate article upon the subject of "Cheap Fuel" in a recent issue of the Chicago Times, we make the following quotations, which appear to be pertinent in this connection, and which, although relating more particularly to the interests of the city of Chicago, are perhaps equally applicable to all that region: —

"The supply of coal from the vast beds which underlie so large a portion of Indiana, Illinois, and Missouri, have gone far to make up for the deficiency of wood in these districts. Yet, notwithstanding these supplies of coal, in many places fuel is both scarce and high. The cost of mining at most of the Illinois mines is, at the present time, not less than $3.50 per ton, while at the mines nearer Chicago, the price is much greater. When we add to this price the cost of transportation and the profits of dealers, it will be found that coal becomes a costly fuel to all who live at a distance from the mines. The freight for one hundred miles on the cheapest route is $2 per ton, and a great part of the Illinois coal brought into Chicago, is taxed from $3 to $5 per ton for freight.

"Illinois, however, furnishes but a fraction of the whole amount of coal consumed in Chicago. During the year ending March 31, 1865, we received 251,038 tons of coal by the lake, most of which was the product of Pennsylvania mines; and but 66,725 tons from the Illinois mines. During the past season, the imports of coal by the lakes increased to 288,771 tons. This coal is carried by railroad from the mines, some of them 200 miles distant, to Lake Erie, and then shipped by water a thousand miles. It will be evident that the cost of this fuel is greatly enhanced by the cost of transportation. So it must always be. Unless such improvements are made in navigation and locomotion as to lessen the cost of transportaion 50 per cent. or more, fuel must be high, except in the vicinity of the place where it is produced.

"These considerations make it important that all the sources of fuel in the west be fully investigated and developed.

"Wood and coal have been long and well known by every one, and wherever found, their value is appreciated; but ignorance of the nature and value of *peat* as a fuel has prevented the development of the beds of this material, which, it seems, are found in many parts of Illinois, Indiana, and Wisconsin, and in some sections where there is no coal, and but very little wood.

"This substance is found, on examination, to be composed of various mosses and aquatic plants, partially decomposed and solidified. It is, in fact, incipient coal, representing the first steps in the progressive changes which transform vegetable substances into mineral coal. Its ultimate elements are the same as those of wood and coal, namely, carbon, hydrogen, oxygen, and nitrogen. According to Sir Humphrey Davy, from 60 to 99 per cent. of the substance of dry peat is combustible, the remainder consisting of earthy matter which forms ashes.

"Machinery has been invented by which the peat, as it comes from the bog, is compressed into hard and compact masses, resembling unburned brick in consistency, and not easily broken. This condensed peat is not only as easily transported as so much coal, but it is also a very superior fuel, burning freely in an open grate, or in a wood or coal stove, with intense heat, and being so compact in form as to present unusual advantages for making steam, especially in locomotive engines, and, indeed, for almost all manufacturing purposes.

"For domestic purposes, it unites the cleanliness of wood with the heating advantages of coal, and for open fires it is charming.

"For manufacturing purposes it has remarkable advantages; for, being entirely free from sulphur, iron can be worked in it without injury. The extraordinary heating power of peat charcoal, together with its freedom from sulphur and other substances deleterious to metals, renders it peculiarly adapted for smelting purposes, and for all manufactures of metals.

"In some countries it is highly prized as a fuel for locomotives, and for making steam generally, for which its strong flame and freedom from clinker admirably qualify it.

"In Northern Illinois, deposits have been discovered in various places; the largest, so far as known, being in Whiteside County, near the Mississippi River. One of these bogs is said to be four miles in length, and one in breadth, and twelve feet deep. The value of these deposits, situated, as they are, where there is very little wood and no coal, is almost beyond computation. Doubtless, peat may be found in many other parts of the State; and it would be well if trials were made of the contents of all the swamps, which, in some sections, render useless a considerable portion of the surface.

"Our farmers may find that these waste places, now producing nothing better than rushes and water-lilies, are the most valuable parts of their land.

"Along Lake Michigan, both in Indiana and Illinois, beds of peat have been found, some of which are of excellent quality, though where the deposits are very near the lake, the quality of the peat has been vitiated by the admixture of sand.

"In view of the importance of the fuel question in the north-west, every effort should be made to discover and develop the beds of peat which are known to exist in many localities, and which probably may be found in many more.

"The wood and coal consumed in the city of Chicago, during the past year, have cost our citizens fully $5,000,000.

"If the peat-beds of the the north-west are once fully developed, it is probable that the cost of fuel will be diminished at least one half, and thus $2,500,000 be saved to the city annually. Moreover, the superior qualities of peat and peat-charcoal for the manufacture of iron and the generation of steam will be sure to secure the attention of manufacturers, and, if we can be abundantly supplied with this material at cheap rates, Chicago will be sure to become a great manufacturing, as well as a commercial city."

Explorations are being made all through this region, and sharp eyes are watching for results; for in no section of the country is there a more ready appreciation of the value of fuel, and wherever peat is found it will doubtless be turned to good account.

Wisconsin.

Throughout the valleys of this State, peat is very abundant; and, in a district where vegetable and mineral fuel is so scarce, it seems highly probable that it will, at some future day, be resorted to as an extremely valuable substitute for coal and wood. These valleys present a very peculiar character in one respect; which is, in the singularly level planes which are maintained in their entire breadth. They appear as if they had once been filled to a uniform level, in the manner of a dam, from bank to bank, or like artificial reservoirs from which the waters have escaped. These level bottoms consist of *peat*-beds to an unknown depth; and small streams meander through them, having muddy bottoms, and frequently expanding into swamps.

It would seem that these Wisconsin valleys have acquired this peculiar uniformity of plane surfaces from the depositions of earthy matter in the first instance, succeeded by the growth and decay of that class of coarse aquatic vegetables which prevail under such circumstances.

Considerable interest was awakened in this subject some ten years ago: the matter was discussed, and efforts were made to utilize the peat for fuel; but, from the fact that no machinery could be procured adapted for the purpose required, no practical results were realized. But now that the needful process of manufacture has been discovered, and machinery effectual for the purpose is to be had, the subject is again receiving the attention it deserves; parties in various parts of the State are investigating it, and the manufacture of peat-fuel has already been commenced in a few places.

We have seen a report made in 1857 by Dr. A. A. Hayes, of Boston, State Assayer of Massachusetts, to some parties owning extensive peat-bogs near Madison, Wis., in which he says of the sample placed in his hands, "Its flaming quality is of a marked character. Its inflammable part has a high heating power, and burns freely and cleanly from the ash. Taking the fifty-nine parts of inflammable compounds as representing the positive combustible matter of this peat, we have a caloric equivalent closely corresponding to that of oak-wood; and I am led by my results to expect an equal heating-power from an equal weight of this peat, burned in comparison with coal." He adds, "Your peat has a marvellous power in producing good gas. It exceeds all common cannels, and, of course, is far above any bituminous coal, and can be worked with *poor* coal to make *good* gas. There are only two or three cannel coals known which afford so much illuminating material, placing this peat in the first class of gas materials."

It is claimed that the most extensive beds of peat known to exist in the west are in Wisconsin.

As this State has but little wood accessible to the fertile and well-settled counties, and no mines of coal, these deposits of fuel are of great importance, and may be of immense value when manufactures shall have been more generally introduced.

From a number of articles upon this subject, which, during the past year, have appeared in the various newspapers of the State, we select one from the "Northern Farmer," published at Fond du Lac, from which to make a few extracts covering a good deal of ground.

"We have been examining into the merits of peat as an article of fuel, and are convinced that it is destined to become one of the most important sources of wealth, as it is in some respects more valuable than coal.

"It can be taken from its bed, and manufactured into fuel, at a much less expense than is required to obtain coal from the most favorably located coal-beds, without taking transportation into consideration. This country contains peat deposits of sufficient magnitude to supply all the wants of a densely-populated community for several generations; so that our valuable timber, now being consumed for fuel, would be preserved for mechanical purposes. We hope that public attention will be aroused to this matter in time to save this wholesale destruction of our valuable oaks, maples, hickory, ironwood, and other varieties of timber. When these are all turned to ashes, necessity will compel us to look for a substitute for fuel: this will be found in *peat;* but we doubt being able to find a substitute to supply our workshops with material that has been consumed to supply our growing population with heat during our long winters.

"Our city alone requires 30,000 cords of wood annually, and the demand constantly increasing. Old residents can see what were extensive forests, rapidly disappearing before the axe; and, unless a check is put upon their wholesale destruction, they will soon be a thing of the past.

"The ingenuity of man now steps forth, and appears in the form of numerous styles of machinery for preparing the products of these heretofore apparently worthless marsh-lands into valuable fuel deposits, each, owing to its ease of access, superior to a mine of coal.

"We believe the mode of preparing consists in pulverizing, pressing, and drying, which is done very rapidly; and in ten days' time it is an article of merchandise ready for use. Samples, that we have seen, appear as solid as Breckenridge coal, and will weigh nearly as much to the cord.

"We should be glad to see some of the surplus money that has been invested in oil-wells pumped out and invested in peat-beds and machinery for preparing it."

Iowa.

The subject of peat is one of vast importance to the people of Iowa, and especially to those in the northern portion of the State, where the article is believed to exist in large quantities, on and near the head-waters of the Iowa and Des Moines. Dr. White, the State Geologist, we are glad to know, is devoting a good share of attention to this subject, among others; and we doubt not his investigations will develop facts which will be highly beneficial to the people of the State.

The great drawback to the settlement of a prairie State, like Iowa, where many miles often intervene without any timber at all, has been the want of fuel; and that will continue to operate against the rapid settlement of the State, and, in fact, of all prairie country, until science shall discover a remedy. The artificial growth of timber has been resorted to, but it is usually for protection only, and not for fuel. Peat and coal, however, offer the readiest solution of the difficulty; and, for that reason, all information upon these subjects, and especially upon the former, which promises to be at once made practicable, cannot but be of value.

The following letter from Dr. White, the State Geologist above referred to, appeared in the "Burlington Hawkeye," in September last: —

"Quincy, Iowa, Sept. 10, 1866.

"Editors Hawkeye:

"Concerning the origin of peat, there is at the present day almost no difference of opinion. It is the result of partial decomposition of vegetable matter under water, which condition arrested the decomposition at a certain stage. Wherever a pond has existed, rank grasses and plants have grown upon its borders, and the frosts of each returning November laid them beneath its surface; their comminuted fragments narrowing the area, and lessening the depth of its waters, until it became the proper habitat of a peculiar moss, which continued to flourish luxuriantly upon the rapidly decomposing bodies of its parent stems, until the pond became filled to the brim with the carbonaceous matter thus produced, resembling in appearance soft black mud, the upper portion being usually thickly interlaced with fibrous roots. Peat, then, can never be found where there have been no ponds, and is not likely to exist, in large quantity, in well-drained regions.

"When the continent was first raised from the diluvial sea, the surface of that portion of it which is now our State, was not so uneven as it now is; but shallow depressions only existed, which gave initial direction to the streams. The rains and frosts of the unnumbered years which have passed since then, have worn their channels deeper and deeper, causing the deepening also of their tributaries, as well as of the multitude of small depressions and ravines which lead into them. Thus, wherever the streams are numerous, and their valleys deep, the country is perfectly drained, and, consequently, no ponds are found. But in a region where many streams have their rise, each depression in the surface would become a pond, because no accumulation of waters beyond sends a current across it to form a channel for its outlet. The southern portions of Iowa present an excellent example of a well-drained region, and a part of the northern portion of the State, near the head-waters of its streams, possesses that character of surface upon which we may expect to find deposits of peat. In some parts of the world this substance forms the chief article of fuel; and the principal reason why its use has been heretofore discarded for that purpose in regions favored with wood and coal, has been the want of knowledge of a proper method of preparing it for convenient use. This is now accomplished so effectually, that, in the Eastern States, peat is coming into successful competition with coal; and, since it contains no sulphur, as bituminous coal usually does, it is considered superior to that fuel for many purposes. Many machines have been constructed for its preparation, some of which have failed to perform the work in the manner desired. The only one of these which has been brought definitely to my notice, and probably the most successful one yet constructed, is manufactured by Messrs. Leavitt & Hunnewell, of Boston, Mass.

"The scarcity of timber in some of the counties of the northern part of the State, and the improbability of finding coal there, make the discovery of peat a subject of the greatest importance. In view of these facts, I have sent my assistant, Mr. Childs, to the counties of Franklin, Wright, Cerro Gordo, Hancock, Worth, and Winnebago, with instructions to examine them for the purpose of ascertaining the extent and character of such deposits there. Peat will doubtless be found in many other parts of the State, wherever the peculiar conditions of its formation have existed; but, considering their position to the streams of the State, the counties named seem to present the best field for present investigations of this kind. C. A. WHITE, *State Geologist.*"

MINNESOTA.

The same general remarks which have been made, as relating to Wisconsin and Iowa, are probably equally applicable, in the main, to this State.

The following, which was cut from the "St. Paul Press," is so much to the point, and tells the story so much more forcibly and concisely than we could expect to, that we are glad to avail ourselves of it, in lieu of a longer article prepared for this edition.

It will be seen that it was written on receipt of a former edition of this work; and while modesty might lead us to clip off the first and last paragraphs, other considerations prevail, and we acknowledge with thanks the kindly notice and manifest good will of the "Press." It says, —

"We have before us a large volume of 168 pages, issued by Leavitt & Hunnewell of Boston, entitled 'Facts about Peat as an Article of Fuel,' in which the whole subject of peat, its origin and composition, its various methods of preparation for fuel, its geographical distribution, its use in the smelting and manufacture of iron, steel, &c., its superiority as a fuel in generating steam for locomotive or mechanical purposes, for the production of gas and for other purposes, are elaborately discussed.

"Peat exists in exhaustless deposits in every northern State, and this volume furnishes a vast deal of valuable information respecting the peat-bogs of each, generally compiled from geological authorities.

"The writer recognizes the existence of large and valuable deposits of peat in Minnesota, but does not seem to be aware that we 'beat all creation' on peat.

"Minnesota is chock full of peat. We don't know how it is in the southern part of the State, but in the northern there is hardly a square mile of land which does not contain a bog of peat.

"There is peat enough in the corporate limits of St.

Paul to supply the city with fuel for many decades to come.

"It is a singular fact that these immense peat deposits, in every northern State, have lain almost entirely neglected, until within the past year or two, — the mass of people being almost wholly ignorant of their existence; while it was only in very rare and exceptional instances that peat was used for fuel anywhere in the United States.

"It is another singular fact, that, by a movement apparently simultaneous and universal, 'Peat' has succeeded 'Oil' as the great excitement of the day, from the Atlantic to the Mississippi.

"The publication of the book before us is itself a curious proof of the wide popular interest taken in this subject. But a still more remarkable and substantial evidence of this interest is manifested in the practical and characteristically Yankee form of 'peat machines,' of which, within a year or two, the mechanical genius of Yankeedom has been prolific.

"We don't know how many varieties of peat machines there are, but the book before us presents an engraving of a set of machinery which is described as capable of turning out 100 tons of crude peat per day, yielding, if cut from a well-drained bog, about 25 to 40 tons of hard, dry fuel; furnished to order for fifteen hundred dollars, — weight about 4500 pounds.

"This is 'Leavitt's Peat Condensing and Moulding Mill,' which we hereby advertise 'free gratis for nothing,' in the hope that some of our enterprising citizens may be induced to give it a trial, and relieve us from the miseries of green bass-wood at eight dollars a cord, or thereabouts."

KANSAS, NEBRASKA, COLORADO, UTAH, AND NEVADA.

Rumors and reports of deposits of peat through this section of the country have reached us, from time to time, but as yet are not of such character as to afford definite or reliable information. Enough has been learned, however, to convince us that there are extensive deposits of good peat scattered through this region.

That such is the fact, has been made so clear to our government that orders have been issued from Washington, to the quartermasters on numerous military outposts and stations, to explore, investigate, and report.

The enormous expense incurred for the fuel required at many of these stations, amounting not unfrequently to fifty, and even one hundred dollars per cord or ton, renders the discovery of local supplies of any material which can be made to serve the purpose, a matter of no inconsiderable importance to our government.

CALIFORNIA.

The *tula* marshes of this region appear, so far as we can learn, to be identical with *peat*. They abound in some sections of the State, and the matter is beginning to be investigated. Government has instituted inquiries and explorations, the result of which will doubtless be made public in due time.

PEAT IN THE MANUFACTURE OF IRON.

It is not generally understood that peat can be advantageously used in the manufacture of iron; but the instances we are able to cite, and the facts we gather from various sources, will, to say the least, go far to show that there is good reason to believe that it may be used economically and profitably, and that it will tend to the production of superior qualities of iron.

The time was, when the English forge-masters maintained, with all the energy of conviction, that pit-coal could never be used in the fabrication of iron; and they treated with ridicule all who made such attempts. We have witnessed the success of those who proved by persistent efforts and experiments that it could be done, and the result is its universal and successful application.

So also with the employment of anthracite coal in the process of iron-making. It had baffled, for a long series of years, every attempt to employ it, and was repeatedly pronounced so surrounded with difficulties as to be impracticable; but it is well known that it is now managed with equal or even more facility than bituminous coal.

We expect to be able to show that it is not only practicable to employ peat as the fuel in fabricating iron, but that it has absolutely been in full operation on an extensive scale; in high furnaces, in puddling and refining, in cubilot and in reverberatory furnaces, in forges, and, in fact, in nearly all the processes of iron manufacture.

It has been satisfactorily experimented upon; but with a very few exceptions, we are not aware that it has been used to any extent, for any of these purposes, in our own country.

One of the largest wire-making establishments in Massachusetts, famed for the superior quality of its wires, especially the finer grades, such as are used for cards, piano-fortes, &c., has consumed annually, for the last twelve years, from 1500 to 2000 tons of peat, cut and dried in the ordinary manner; and the quality of the wire produced is said to be owing, in great measure, to the use of this fuel.

In England, France, Italy, Bohemia, Bavaria, Westphalia, Wurtemberg, and in several adjacent provinces, it has been successfully employed on a large scale, and with very satisfactory results. These practical tests and proofs of the value of peat for the purposes under consideration, surely cannot, if generally known, remain long unheeded or unimproved by the enterprising iron manufacturers of our own country.

In a paper on this subject, by Mr. A. S. Byrne, published in 1841, he remarks that charcoal iron is the best known at present in the markets, and, as an illustration of its value and superiority, states that large quantities had been annually imported into England from India and China, and sold at the (then) enormous price of £36 (=$173) per ton. He contends that *peat-coke is of still greater value than the best charcoal*, and that *in the manufacture of iron it stands unrivalled as a fuel.*

The admixture of peat, even in its natural state, with common coke, in smelting iron, materially improves its quality; in some instances changing the pig-metal from the state of "white iron" to that of "gray iron," technically called "foundery."

Good peat is shown to be preferable to any other fuel, not only for the process just mentioned, but in welding, and for softening steel plates, &c.

For the finer iron-works, peat and peat-charcoal are known to be better than wood-charcoal.

Sir Robert Kane, in his "Industrial Resources of Ireland," published in 1844, demonstrates that the precious Baltic iron, for which, at that time, £15 to £35 per ton was readily paid, could be equalled by Irish iron, smelted by Irish turf, for £6 6s. per ton.

From other sources we learn that iron manufactured

with peat fuel is more malleable than Swedish, and that tools made from it are of superior quality.

It has been doubted if peat could be used in the puddling furnace, except with a diminished produce; yet the working of iron by this fuel is known to improve its quality, and the welds, especially, are superior to those made with coal.

It has been proved that, after peat has been well carbonized, it may be employed in puddling and reverberatory furnaces and forges. As to its use in blast furnaces, peat, which is the lightest of all coals, would consequently seem to be the least fitted for the reduction of ores. But even this difficulty has, in great measure, been surmounted in the high furnaces of Germany.

M. V. Lamy made a series of experiments to determine the quantity of heat evolved by the burning of peat, compared with other combustibles, with the following results: —

One kilogram, or 2¼ lbs., of the varieties of fuel mentioned below, evolved of caloric the following parts: —

Wood charcoal	75	parts
Coal Coke	66	"
Charred Peat	63	"
Bituminous Coal	60	"
Charred Wood	39	"
Dry Wood	36	"
Raw Peat	25 to 30	"
Wood with ¼ moisture	27	"

Thus, as regards charred peat or turf charcoal (from *unsolidified* peat), it appears preferable to bituminous coal in the manufacture of iron, and is almost equal to wood charcoal.

Compressed peat, thoroughly dried, gives a steady

and intense heat, and can be used with decided advantage in a puddling furnace. In fact, it has been extensively used, and the results have been very carefully investigated by men of science, as well as of profound practical attainments, and with uniformly gratifying results, demonstrating conclusively the superiority of the article, although it cannot be claimed that any one has yet arrived at a perfect method of preparing and using it; but enterprise and inventive genius are fast developing both.

From a statement of experiments made by M. Le Serge, found in the "Repertory of Arts," vol. v., it would appear that ordinary turf, charred, is capable of producing a far more intense heat than common charcoal. It has been found preferable to all other fuel for case-hardening iron, tempering steel, forging horseshoes, and welding gun-barrels.

Since turf is partially carbonized in its native state, it must evidently, when properly manufactured and condensed, and fully charred, afford a charcoal very superior in calorific power to the comparatively porous article prepared from wood by fire.

At Königsbronn, in Wurtemberg, they execute with peat alone, the refining and second fusion of the pig metal; its puddling, the reheating of the lumps and rolling the bars and plates; in fine, all the operations that are made with coal in English forges. The works are under the care of M. Veberling.

The peat is of three kinds, as follows: —

1st. Peat of Dattenhausen. — Color varying from dark yellow to brown, and containing a very considerable amount of fibre or interlaced filaments; yields ashes 3½ to 4 per cent.

12

2d. Peat of Guntzburg. — Color, dark brown to black, compact, and yields ashes 6 to 7 per cent.

3d. Peat of Wilhelmsfield. — Dark brown, and yields ashes 5 to 6 per cent.

The peat used at these works is first dried in the air at the place where it is dug, the blocks being turned occasionally, and, after some three or four weeks' exposure, are transported to the iron-works, and there further dried in kilns by artificial heat, and stored in dry quarters for use.

Of the three kinds of peat mentioned above, the proportionate diminution of weight and volume, when dried, is as follows : —

	1st.	2d.	3d.
Diminution of volume . . .	.24	.10	.13.5
" " weight. . . .	.10	.19	.12

Cost of one metrical quintal = 220 lbs., delivered at the iron-works of Itzelberg, is 1 fr. 29 c. = 36 cents, or about three dollars and fifty cents per ton; the distance being two kilometres, or 1⅓ miles.

M. Berthier's analysis of peat used at Königsbronn is as follows : —

Carbon	24.40
Volatile Matters	70.60
Ashes	5.

It is employed without admixture of other fuel in the refining, puddling, and reverberatory furnaces.

According to M. Lefebvre, whose statements appeared in the "Annales des Mines" in 1839, peat was used to a considerable extent at the iron-works of Ichoux, in Les Landes; and the proportions which resulted from the operations at the refining and puddling furnaces and forge operations at these works, chiefly through the use of peat, were as follows : —

114 kilograms pig iron produced 100 kilograms bar iron, with 93 kilograms peat, and 52 kilograms wood.

116 kilograms pig iron produced 100 kilograms bar iron, with 93 kilograms peat, 37 kilograms wood, and 9 kilograms coal.

The peat of Ichoux contains two and a half times more ashes than the peat used at Königsbronn, in Wurtemberg, of which mention has just been made on previous pages.

In 1842, the establishment in Ichoux was the only one in France where peat was used to any extent for the processes above referred to; but with the knowledge of the success attained there, and the low price at which it might be obtained at numerous points in the kingdom, it would seem impossible that many years should elapse without its general introduction and use in the numerous iron and steel works there.

At the iron-works of Ransko, in Bohemia, peat is used with great success. These works are situated at the south-west extremity of Bohemia. They consist of two high furnaces and two cubilots, which are worked with a mixture of turf and charcoal. There are also several refining fires, and the establishment gives employment to about four hundred men.

The turf is brought from turbaries situated some leagues from Ransko. It is there dug or cut in the usual manner, in bricks or oblong pieces, about fourteen inches long by six inches square. These bricks are exposed in piles in the open air during the fine season, where, in drying, they contract to nearly one third their original size. In general these peat bricks are not used until a year after having been dug. They are stored under sheds attached to the high furnaces, and sheltered from the rain, but receive no other or further attention or preparation. It was at first proposed to use it in the carbonized state; but, as regards this particular kind of peat, the carbon obtained was not found to be much, if any, more advantageous than the article in its crude state; and as the process involved some expense, without corresponding profit, it was abandoned as useless. Numerous other experiments were tried in the preparation and curing of it; but most were abandoned, for the reason that little, if any, profit was obtained over the original methods adopted; and they therefore continued to employ non-compressed turf, simply dried in the open air.

Two varieties of peat are used here, weighing respectively four hundred and five hundred and eighty-seven lbs., the cubic metre of thirty-five and one quarter cubic feet English. They cost at the iron-works 13d. English, or 26c. United States, per cubic metre of thirty-five and one quarter cubic feet English. The weight and cost of the same measurement of the different kinds of charcoal used in the high furnaces with the peat are as follows: —

	Weight.	Cost at the Works.
Charcoal, resinous wood . .	275 lbs.	.80c.
" heavy wood . . .	468 "	1.06c.
" as mostly used . .	314 "	.84c.

The cost of a volume of charcoal is thus shown to be more than three times that of an equal volume of peat, and it would therefore seem to be desirable to exchange, as soon as possible, the charcoal for the peat.

The ore smelted here is clay iron-stone of moderate quality, and the fuel is generally turf and charcoal mixed. The quantities employed in the making of a ton of iron are, turf thirty-five cwt., costing 8s. 10d., and charcoal thirty cwt., costing 24s. 7d., — together,

£1 13s. 5d.; and iron of the very highest character is produced.

At Schlakenwerth, in Bohemia, near Carlsbad, is a high furnace which works with a mixture of equal quantities of wood charcoal and peat charcoal. The peat is found upon the plateaux of the Erzgebirge, at an elevation of more than one thousand metres, and the weather and climate are such that it can be produced to advantage only during about two months in each year. It is then carbonized in much the same manner as wood, in circular piles; and a very dense charcoal is obtained, which on an average does not contain more than five per cent. of ashes. The cubic metre of thirty-five and one quarter cubic feet English of this peat charcoal weighs six hundred and sixty lbs., while the same amount of wood charcoal weighs only three hundred and ten lbs. The analysis of this peat charcoal shows fixed carbon sixty-seven, volatile matters thirty, ashes three.

In the cubilot furnaces of Bohemia, also, a mixture of equal parts of peat charcoal and wood charcoal is employed with results highly satisfactory.

Peat is employed in the iron-works of Weiherhammer, in Bavaria. It is procured from the numerous tourbieres of the Fitchtelgebirge, which are worked during the fine season, after which the turf is left to dry for about six months. It is then stored, but is not employed in the iron-works until a year after it has been dug. The peat is of good quality, compact and heavy, and, on burning, yields not more than four to five per cent. of ashes.

At the Weiherhammer Works are two puddling furnaces, one of which is generally in operation. The puddled iron is converted into bars in the ordinary charcoal forges, or in a chafing fire, which is fed with peat alone. The density of this peat is hardly sufficient to produce the temperature required to remelt the iron; combustion is therefore hastened by means of a forced current of air, which is furnished by the blowing machine of the refining furnace, and in this manner the re-melting of the pig metal is effected with the greatest facility. The result of these operations is stated as follows: Two hundred and eighty-one lbs. pig metal, with eighty-five and thirty-two one-hundredths cubic feet raw peat, produced two hundred and twenty lbs. bar iron. These proportions are equivalent to two thousand eight hundred and sixty-one lbs. pig metal and eight hundred and sixty-eight cubic feet of raw peat to produce one ton (two thousand two hundred and forty lbs.) of bar iron.

In Germany, the gas of the high furnaces where peat is used has been satisfactorily employed in the refining of iron and the puddling of steel. At Magdeburg, in the Hartz, not only is iron refined, but steel of excellent quality is fabricated, by the use of gas. The beneficial results obtained by the use of gas in refining iron, the economy of the fuel, the smallness of the loss, and the amelioration of the quality, have been urged as reasons why it should be extensively introduced in peat-producing countries.

It is considered that the known successful results of this method are of the greatest importance to the whole of Northern Germany, where extensive beds of turf and lignite prevail, which will afford a great resource to those districts. The same process would not be less beneficial to France. It has been adopted to a considerable extent in Sweden, where coal is scarce.

In an able pamphlet published some twenty years ago by J. W. Rogers, it was suggested that the overplus working population of Ireland might be permanently and advantageously employed in the preparation of different kinds of fuel from the immense bog districts. He states that he "has been in the habit of having peat charcoal prepared for smith's use, infinitely in preference to any coal," and that "if within the reach of the manufactories of iron, at the price for which it can be produced, *no other fuel would be used*." He adds, "Charcoal of peat has been found by analysis to possess almost identical qualities with wood charcoal. Prepared as it hitherto has been, however, it is more friable, and therefore more fitted for many purposes, such as the working of iron, manufacture of gunpowder, &c.; but peat charcoal is quite capable of being prepared so as to obtain a density little if at all inferior to wood charcoal." This, it should be borne in mind, refers to charcoal from peat in its crude or unmanufactured state. Mr. Rogers adds, — and the correctness of his remarks is confirmed by statements from numerous sources, — that, "when *condensed* peat is carbonized, it gives a fine coherent coke, which amounts to about thirty per cent. of its weight, and contains very little ash. The density of this coke is *greater* than that of wood charcoal, being found to range from 913 to 1040."

The objection often urged, as regards iron ships of war in action, that the splintering is so great that this material cannot be safely used, is met by the assertion that the evil arises entirely from one cause, — that of iron being made mostly by sulphurous fuel. *Iron made with peat charcoal will not splinter.*

We have seen it stated, that, in some parts of England, horse-shoes made with peat are considered so much more durable than those made with any other fuel, that nearly double the ordinary price is paid for them.

M. Bussou de Maurier, *à propos* of the prizes of three thousand francs, offered by the Société d'Encouragement of Paris, in 1855, for the best process by which a fuel adapted for household and manufacturing purposes may be economically prepared from peat, notified that society to the effect that he had succeeded in preparing an excellent solid, compact, and tenacious charcoal, or rather coke, by distilling peat mixed with small bituminous coal.

This coke, he says, is admirably adapted for the forging of steel and other metallurgical operations.

A paper of considerable interest and importance appeared in the "London Times," soon after the International Exhibition of 1862, upon the distinct varieties of fuel, peat, and coal, of which several specimens were there exhibited.

Not much was to be learned from the display of mere lumps of raw peat; yet there were several illustrations connected with peat, of great practical value.

Peat is applied as fuel in Great Britain in the smelting of lead in the "ore hearth." This method is extensively employed in the north, and, in the case of rich ores, is maintained by some experienced smelters to be superior to every other.

It is a singular circumstance, that in former times it was practised successfully in Derbyshire, where it became extinct; and that attempts to re-introduee it have signally failed.

In Bavaria, peat-bricks are extensively used under locomotive boilers; and, in Sweden, Ekman has long employed peat in his gas-welding or reheating furnaces. It is fashioned into bricks, and subsequently dried by artificial heat, at a temperature almost sufficient to cause incipient charring.

In the Swedish department were specimens of iron manufactured in such furnaces, with prepared peat as the fuel, by Baron Hamilton, Nericia, at whose works the annual consumption of peat for this purpose is said to be very large.

In the Italian department, And. Gregorini exhibited steel made in Siemens' gas puddling furnace, with peat and lignite as fuel.

Furnaces constructed on this principle, whether for the use of peat or other kinds of fuel, are strongly recommended as worthy the attention of our iron-masters and other practical metallurgists who are interested in economizing fuel. It is claimed also that they are specially adapted for the manufacture of glass.

Messrs. Siemens, of London, are said to have been most successful in the application of this principle; and we have, therefore, obtained from their agents, Messrs. Tuttle, Gaffield & Co., of Boston, a description of their furnace, which we insert at length, not only for the interest which we are sure it will have with iron and glass manufacturers generally, but from the fact, so distinctly claimed for it, that it is especially adapted for the successful and economical treatment of metals and glass, by the use of *peat-fuel.*

The Siemens Regenerative Gas Furnace is the joint invention of Messrs. C. W. and F. Siemens, of London. For many years they labored to bring their ideas into a thoroughly practical form; and the result has been the introduction of one of the most perfect arrangements for the conservation and utilization of heat that has ever been put in practice.

The inventors claim the following advantages for the Regenerative Gas Furnace.

1st. Saving of fuel, amounting to from forty to fifty

per cent. in the quantity, besides which the most inexpensive qualities of fuel, such as peat, slack, coke-dust, and lignite may be employed, producing a money-saving, in many instances amounting to seventy-five per cent.

2d. Unlimited command of heat without intense chimney draught, owing to the principle of accumulation involved.

3d. Great purity and gentleness of flame, which largely diminishes the oxidation or deterioration of the material heated in the furnace, and improves the quality of the product.

4th. Increased durability of furnace, owing to the absence of ashes, and a perfect uniformity of heat throughout the furnace.

5th. Saving of space within the works, and great cleanliness of operation, the fuel being converted into gas outside the works.

6th. Complete command of the intensity of the heat, and of the chemical nature of the flame, which may be arrested or changed from a reducing to an oxidizing flame, or the reverse, at any one moment, tending to facilitate and improve all metallurgical operations.

7th. Complete absence of smoke from the stack, which renders this furnace beneficial to the public in large towns.

As now made, this furnace consists of two distinct parts: the producer, in which the fuel is converted into gas for supplying the furnace, and the furnace proper, including the regenerators.

The gas-producers are of various forms, according to the nature of the fuel used, and are quite simple in their construction. They are situated outside the furnace building, and the gas is conducted to the furnace, through flues, to any distance which may be desired.

The furnace is so constructed that, underneath the heating chamber, are placed transversely four regenerator chambers, which are filled with fire-bricks, built up with spaces between them to admit the passage of gas and air.

These regenerator chambers work in pairs, the two under the right hand end of the furnace communicating with that end of the heating chamber, while the other two communicate with the opposite end. For instance, the gas from the producer passes through the main gas flue, and enters at the bottom of one of the regenerator chambers, at the right-hand end of the furnace, and the air at the bottom of the other, whereby they are kept separate up to the moment of entering the heating chamber, but are then able to mingle intimately, producing at once an intense and uniform flame, which distributes itself all over the heating chamber. The heat, having once performed its work in the furnace, is not now wholly released from further service, or only partially utilized, as in the old style of furnaces, but is bound to an endless round of duty, and is carried down into the other pair of regenerators, and all but a very trifling portion of it is there arrested by the packing of fire-bricks, and garnered up for future service.

When this pair of regenerators has become fully heated by the passage of the hot products of combustion from the furnace, and the opposite pair correspondingly cooled by the upward passage of the cold gas and air, the valves in the main gas and air flues are reversed, and the separate currents of gas and air enter at the bottom of these heated regenerators, beneath the left-hand end of the furnace, take up the heat that has been stored in them, carry it back to the furnace, and the surplus is then deposited in the right-hand set of regenerators. And thus the process of reversing is repeated at fixed intervals, so that two of the regenerators are always accumulating the heat which would otherwise be wasted, but which, when the action of the furnace is reversed, is always carried back to the furnace by the incoming currents of gas and air, which, as they pass upwards through these heated regenerators, attain a temperature equal to a white heat, before they meet and ignite in the furnace, and thus add the carried heat to that due to their mutual chemical action. The arrangement is so perfect that, while the furnace and upper portions of one set of the regenerators are charged with a most intense heat, the amount which escapes through the waste flue will not exceed 300° Fahr.

Although the saving of heat is made most prominent in this description, there are other great advantages gained by this process which would commend this kind of furnace to general use even if the same amount of fuel was required to work it as in the old process.

It is evident that the purity of the flame not only insures, in all metallurgical operations, a superior product, but a much smaller percentage of waste, and an increased durability of furnace. And the pecuniary value of these combined advantages far exceed even the gain which is made in the cost of fuel.

One of the most important advantages to be gained by the adoption of this invention is observed in the fact that it utilizes what has been heretofore regarded as inferior fuel, such as sulphurous coal, and the slack and waste of the mines, &c., as also those immense deposits of *peat* to which the attention of manufacturers is now being largely directed. Professor E. Daniels, of Chicago, in alluding to this subject, says, —

"It is scarcely possible to estimate the advantages

which the west may derive from this invention. Our coal is at once redeemed from comparative inferiority, and becomes a source of cheap heat and motive power, adequate, both in quantity and quality, to the vast and varied demands of manufacturing industry."

What Professor Daniels so truly says of the inferior coals of the west, will apply with tenfold force to those extensive tracts of undeveloped fuel which abound throughout the northerly portion of the United States. This fact has been recognized in Europe; and in many parts of the Continent, particularly in Italy, where coal is comparatively dear, *peat* has been used in these furnaces for the manufacture of iron and steel, with the most successful results; and in this country measures are now being taken to build furnaces in which peat alone shall be used, on the plan of those now so successfully worked by M. Gregorini, in Lombardy.

After years of patient application, the Messrs. Siemens now have the pleasure of seeing their valuable invention adopted and successfully operated by the most important of the iron, steel, and glass manufacturers in Europe. Among them can be enumerated the establishments of Sir W. G. Armstrong & Co., Messrs. James Russel & Co., Mersey Steel and Iron Co., London and North-western Railway Co., Messrs. T. Firth & Sons, Messrs. Nayler, Vickers, & Co., the Royal Arsenal, Woolwich; Messrs. Krupp & Co., Essen, Prussia; Messrs. Mayer & Co., Styria, Austria; Arsenal Imperial de Lorient, France; Peelroso Iron Co., Seville; M. Gregorini, Italy; the Paris Gas Co.; the Plate Glass Co.'s of St. Gobain, Cixey, Aix la Chapelle, &c.; the Vielle Montagne Zinc Co., &c.

It is natural to suppose that the manufacturers of the United States will not be backward in securing the use of this important invention; and several of the most enterprising iron, steel, and copper manufacturers have already arranged for the immediate construction of these furnaces in their works: and the results which these will show cannot fail to insure their universal adoption, and must produce a complete revolution in the application of heat in all metallurgical operations.

The Moulded Peat Charcoal Company, at the International Exhibition of 1862, "showed a material of apparently excellent quality; but, without satisfactory statistics on the economy of the processes by which it was produced, the jury could not pronounce upon it."

The jury (1862) gave an honorable mention to Mr. J. Brunton. "His statements as to employment of the condensed peat on a large scale referred only to a very short period, and were considered as requiring further confirmation."

Mr. W. E. Newton, in 1862, before the Society of Arts, said that peat, if properly prepared and properly used, gave a calorific power equal to, if not greater than coal; but the use of peat in manufactures was of greater importance than simply as a fuel for heating purposes. A great many experiments, more particularly on the Continent, had been made with peat as a fuel for metallurgic purposes, and it had been found that it produced iron of a very superior quality. He had seen specimens which came up to the best quality of Swedish iron. Every iron manufacturer knew that if he could get peat to stand the blast, then it was infinitely superior to coal for those purposes, for the simple reason that it contained no sulphur. They could produce iron with peat, from the worst brands, which would almost equal the best Swedish or Russian iron, simply owing to the absence of these deteriorating chemical agents which existed in coal.

D. K. Clark, C. E., in a paper read before the British Association in London, in 1865, says,—

"In Germany, peat mixed with wood-charcoal is very extensively used in the production of iron; the peat, as prepared there, not being sufficiently solid to do the work alone: but it is found that the greater the proportion of peat that can be used, the better is the quality of the iron produced. The gas delivered from the high furnaces has also been satisfactorily employed in the refining of iron and the puddling of steel. The value of peat in the production of iron has long been established. Iron metallurgists are agreed in the opinion that iron so produced is of very superior quality. In every stage of iron manufacture, and in welding, peat-charcoal is most valuable. At Messrs. Hick & Son's forge, in Bolton, a large mass of iron, about ten inches square, was heated to a welding heat with peat charcoal, made at Horwich. The time occupied was less than the operation would have taken with coal; the whole mass was equally heated through without the slightest trace of burning on the outside; and in hammering out the mass, as much was done with one heating as ordinarily required two heatings to effect. The importance of obtaining an abundant supply, at cheap rates, of peat-charcoal, cannot, therefore, be too highly estimated."

The charcoal made from peat, at Horwich, is extremely dense and pure. Its heating and resisting powers have been amply and severely tested, and with the most satisfactory results. At the Horwich Works pig-iron has been readily melted in a cupola. About 80 tons of superior iron have been made with it in a small blast furnace, measuring only six feet in the boshes, and about 26 feet high. The ore smelted was partly red hematite, and partly Staffordshire; and the quantity of charcoal consumed was 1 ton 11 cwt. to the ton of iron made; but, in a larger and better-constructed furnace, considerably less charcoal will be required. It has also been tried in puddling and air furnaces, with equally good results, considerably improving the quality of the iron melted. For this purpose, the fuel was only partially charred, in order not to deprive it of its flame, which is considerably longer than that from coal. Some of the pig-iron made at Horwich was then converted into bars, which were afterwards bent completely double, when cold, without exhibiting a single flaw.

Mr. Sanderson, of Sheffield, in a report made to an English company, says, —

"You are certainly the first who have succeeded in smelting iron ores in a blast furnace with peat *alone* as a fuel; and I am convinced that you have thereby opened out a large field of industry both for England and Ireland, as well as Scotland, and doubtless other countries.

"All iron metallurgists have agreed in one opinion, that if peat by any means could be produced of sufficient *density* to enable it, when charred, to stand the blast necessary for the production of iron, the iron so produced would be of a very superior quality. Some have thought it would be even superior to iron made with wood-charcoal. You have most fully succeeded in obtaining this density in your peat fuel. Indeed, I have seen samples, taken from the bottom of your furnace, which had been subjected to a high heat from forty to fifty hours, perfectly hard and strong. The sample of pig-iron which I have received and tested for malleable iron, is very satisfactory. As you progress, and pro-

produce grayer pig-iron, and further manufacture it into malleable iron, its superior quality will be appreciated for cables, boiler-plates, armor-plates, wire, and all other kinds of iron requiring more than ordinary strength. Do not allow yourselves to suppose that you have already produced iron of the value and quality which you will be able to produce when proper appliances are brought to bear upon what you have already done."

Mr. Fothergill, when reporting on experiments with peat-charcoal iron conducted under his supervision at Messrs. Platt's Iron Works, at Oldham, says, —

"I have no hesitation in stating that the experiments were a great success. The directors can judge of the tenacity and quality of the iron from the severe character of the test to which it has been submitted, namely, having been completely doubled over *when cold*, without exhibiting a single crack. I deem it my duty to assure the directors that I am fully aware of the important position in which they are placed; yet I have no hesitation in saying, that I have every confidence in their ultimate success: and the opinions I have previously expressed as to the importance and value of the undertaking remain not only unchanged, but considerably strengthened, by the results of the experiments referred to."

Messrs. Brown and Lennox certify, that the strength of the peat-charcoal iron, proved by them in the ordinary manner, is considerably above the average strength of iron of the best brands. The latter testimonial is especially valuable, considering the disadvantage necessarily attendant on smelting iron in a new furnace with new materials.

Professor Emmons, Geologist to the State of New

York, concludes some remarks on the subject of peat as follows: "I shall state only one more application of this material, viz., as a substitute for charcoal in the reduction of iron. The coal which is formed from it is equal to any coal: hence it may become of great importance in those sections of country where fuel is scarce, or as it furnishes a resource in this important business when the ordinary means are expended."

Mr. McDougall, of the Caledonia Iron Works, Montreal, who supplies the Grand Trunk Railway with car wheels, states that, for giving toughness to the metal and uniformity of chill, qualities so essential to car wheels, peat fuel is unsurpassed.

We have the following brief report of an experiment in smelting iron with peat, at these works, made in October last. The cupola was charged with two layers of iron and anthracite coal. The third or topmost layer was iron and peat. The time was forty minutes less than with coals alone. The iron smelted by the peat was hotter when drawn off from the coals, and was said to be more compact, and more like wrought iron, than the other. The test was a severe one, the proportion being twelve of iron to one of peat; the proportion for coals being seven to one.

The "Montreal Gazette," of Dec. 1, 1866, says, —

"We were shown, yesterday, a small piece of bar-iron from the puddling and rolling mills of Messrs. Morland, Watson, & Co., from the first blooms ever made in this country with peat fuel alone, and, we believe, the first on this continent. The specimen shown to us was of the very highest quality, and equal to the very best Swedish iron. It was bent, when cold, by a vice, and doubled close up at right angles, with an edge without a crack or flaw appearing, the outer edge remaining smooth and sharp. A severer test of the tenacity of the iron could not have been applied; a result so satisfactory could scarcely have been hoped. We are told that no iron manufactured from coal in this country would stand such a test. The fact is one of great importance for Canada, in view of its large supplies of peat and iron. We may add that the time taken in the manufacture was not greater than that usually taken when coal is used. There was no special adaptation of appliances. The furnace was an ordinary coal one, and the men were accustomed to the use of coal."

Mr. Campbell, manager of the mills above referred to, states that —

"The peat fuel was tested in an ordinary puddling coal furnace, and no alteration or adaptation was made, although this might have been done, and a large saving of fuel effected.

"The pig-iron used was Dalmellington brand A, a strong iron, soft and very tough.

"The quantity of peat fuel consumed was nearly double the weight of coal used on ordinary occasions.

"In my opinion, and with the present furnaces, by mixing peat with Pictou coal, we could produce iron equal to the best charcoal iron, and at no more expense than the present cost of our iron, the quality of which is equal to the best refined English iron.

"With the furnaces as at present constructed we could not use peat alone, the combustion of the gas given out not being sufficiently perfect to produce the heat required for puddling to advantage, resulting in waste of fuel, and additional labor to the men.

"If we could get the extra price for the quality of iron turned out, there would be no doubt about the result, but I fear this could not be obtained, as almost any description of iron seems to sell which is cheap.

"I send you samples of the iron, made at the trial, which I consider equal in quality to *best charcoal iron*, and superior, almost, to any description of iron imported."

Professor Johnson remarks, "Dried peat is extensively used in puddling furnaces, especially in the so-called gas puddling furnaces, in Corinthia, Steyermark, Silesia, Bavaria, Wirkenberg, Sweden, and other parts of Europe. In Steyermark, peat has been thus employed for twenty-five years.

"When peat or peat coal is employed in smelting, it must be as free as possible from ash, because the ash usually consists largely of silica, and this must be worked off by flux. If the ash be carbonate of lime, it will, in most cases, serve itself usefully as flux. In hearth puddling, it is important, not only that the peat or peat coal contain little ash, but especially that the ash be as free as possible from sulphates and phosphates, which act so deleteriously on the metal.

"It has been found, at Rothburga, in Austria, that by substitution of machine-made and kiln-dried peat for wood in the gas puddling furnace, a saving of 50 per cent. in the cost of bar-iron was effected in 1860.

"What is to the point, in estimating the economy of peat, is the fact that, while 62 cubic feet of dry fir-wood were required to produce 100 lbs. of crude bar, this quantity of iron could be puddled with 43 cubic feet of peat.

"In the gas furnace, a second blast of air is thrown into the flame, effecting its complete combustion. Dellvik asserts, that at Lesjoeforss, in Sweden, 100 lbs. of kiln-dried peat are equal to 197 lbs. of kiln-dried wood in heavy forging.

"In other metallurgical and manufacturing operations, where flame is required, it is obvious that peat can be employed."

A gentleman formerly employed in the iron business in Austria, writes to us recently, "Peat is the only fuel of many rolling mills in Austria. The peat of the Alps there is of good quality. In a puddling furnace there are wanted 163 to 200 English cubic feet of dry peat per English ton of puddling-iron, and in a reheating furnace about the same quantity of very dry peat per ton of wrought iron. This iron is puddled from white pig-iron. The furnaces for puddling and reheating are all gas furnaces.

"Also in blast furnaces for making pig-iron, peat is used in Austria, but not converted into coal, only dry. Peat is mostly mixed with charcoal in the older Austrian blast furnaces, the half of every part; but in blast furnaces, which are built especially for using peat, this fuel can be used alone or mixed with coke or anthracite."

The "London Mechanics' Magazine," in a recent issue, after dwelling at some length on the use of peat fuel for railroad and steamship service, concludes as follows: "Mere compression of the peat is not sufficient to insure its economical use: it requires *condensation*, which quality cannot be imparted to it by the most powerful pressure.

"But however successful may be the results of using peat, either in locomotives or for marine boilers, we do not think it is destined, at present, to play so important a part here as in the manufacture of iron. It is here that peat fuel will, in all probability, come first into exclusive use, owing to its great superiority over coal in every stage of iron and steel manufacture."

PEAT AS APPLIED FOR GENERATING STEAM.

Sir R. Kane, before quoted in these pages, stated, in 1844, that the steamers plying between Limerick, Clare, and Kilrush, in Ireland, were using peat for fuel. The "Garry Owen" steamer made the passage between Kilrush and Limerick, fired with turf (although in the midst of a coal region), in three hours twenty minutes: distance, about forty-five miles. The Shannon steamers were mostly using it; and its consumption in mills, factories, workshops, as well as for domestic purposes, was steadily increasing.

Some years since a patent was obtained by Mr. Williams, managing director of the Dublin Steam Navigation Company, for a method of converting the lightest and purest beds of peat moss or bog into the four following products, each of which possesses very valuable properties:—

1. A brown combustible, solid, denser than oak.
2. A charcoal twice as compact as that of hard wood.
3. A factitious coal.
4. A factitious coke.

One of the most important results ascertained with these products was, that with ten hundred weight of the factitious coal the same steam-power was obtained in navigating the company's ships as with seventeen and a half hundred weight of pit-coal alone: thereby saving thirty per cent. in the *stowage* of fuel.

Mr. D'Ernst, artificer of fireworks at Vauxhall, proved, by the severe test of colored fires, that the turf-charcoal of Mr. Williams was twenty per cent. more combustible than that of oak.

Mr. Oldham, Engineer of the Bank of England, applied it in softening his steel plates and dies, with remarkable success. Mr. Williams's method of preparing the peat was given as follows: When freshly cut, the fibre of the peat is broken up, and the mass is placed between cloths, and pressed by a powerful hydraulic press, which condenses it to one third of its original volume, and to three fifths of its weight, through the loss of moisture. This condensed peat, when carbonized, gives a fine coherent coke of about three tenths the weight of the turf as cut; burns freely, producing intense heat, and leaving only a very small percentage of ash. Its density is greater than wood-charcoal, and the cost of production is about twenty shillings per ton.

Mr. Brunton, before the Society of Arts, in 1862, called attention to the fact, that trials with his prepared peat, upon a sufficient scale, under the boiler of an ordinary steam-engine of twenty-horse power, had demonstrated that the peat, as a heating power, did *two and one third times* the duty of coal. The ordinary consumption of coal in the furnace, was twelve cwt. per hour. An equal quantity of his prepared peat lasted two hours and twenty minutes, producing the same amount of steam per hour, and doing the ordinary work of the engine.

Mr. Paul, in the paper we have before referred to, says, "As regards the use of peat for fuel, it now remains only to consider what are the circumstances under which it can be used for this purpose, and under which there is an advantage in using it rather than coal.

"I can best illustrate this by a case within my own experience. During the last four years I have had occasion to manufacture a large quantity of bricks in one of the western islands of Scotland, and for that purpose required fuel for raising steam to drive the brick machinery, and for burning the bricks. Coal could be delivered at the works, including 4s. for cartage, at 22s. per ton: but I found that the peat, of which there was abundance close to the works, was capable of raising steam well, and of being used for burning the bricks, and that I could, for 8s., put down at the boiler, or at the kiln, a quantity of peat equivalent to one ton of coal; thus making a difference of 14s. between the use of a ton of coal and the use of a quantity of peat equivalent to it.

"This was equivalent also to a saving of 7s. per thousand in the cost of the bricks. The advantage would have been still greater had there been more efficient means for bringing in the peat from the moor, which, as it was in this case, cost as much as the peat on the moor, or about 2s. per ton.

"The applicability of peat for the purpose of fuel on board steamers is indubitable. I have employed peat as the only fuel for steam-boilers during the last four years, and have found it to answer admirably. It has also been tried by Mr. James Napier, of Glasgow, on board his steamer, the 'Lancefield;' and he is of opinion that it might be used in the place of coal."

Professor Emmons, in a report to W. H. Seward, Governor of New York, in 1839, says, "I have been informed that peat, as a fuel for steam-engines, has been proved, by actual experiment, of great value. To impart to it the power of emitting, during combustion, a lively flame, a small quantity of tar is mixed with it, which, of course, creates a larger volume of flame, which is a matter of considerable moment when employed in generating steam. The experiments referred to were made on board the 'Great Western' during her last passage; and such was the result, that a large amount of peat was taken on board for her homeward passage."

From the "London Engineer" we learn that "a paper was read by Mr. P. F. Nursey, before the Society of Engineers, at Exeter Hall, which contained some interesting and valuable information. The paper alluded to the probability of an exhaustion of English coal-fields, owing to the increasing annual consumption, which, between the years of 1850 and 1860, was at the rate of two and three fourths million tons. . . . The question of *peat* was considered at some length. The author, considering the subject of great importance, had collected much valuable information.

"The deposits of peat in Great Britain and Ireland were stated to occupy an area of about six million acres, and to vary in thickness from two feet to fifty feet; and, at an assumed average of twelve feet, an acre would yield about 3500 tons of dried peat, or a total quantity of 21,000,000,000 tons. The process of manufacture of 'condensed peat' was fully detailed. It was shown that the cost of the fuel produced did not exceed that of coal at the pit's mouth. It was shown to possess qualities superior to coal, especially its heating power. Trial of the condensed peat had been made by Mr. B. Fothergill on a river-steamboat, in which twelve cwt. were consumed in two hours twenty minutes, the ordinary consumption of coal being twelve cwt. per hour. The peat gave no smoke, and left no clinkers. The locomotive engineer of the Belfast and Northern Counties Railway had tried the condensed peat on that line. In a trip of seventy-four miles, the total quantity of fuel burnt was fourteen cwt., one quarter, fourteen lbs., — the train, including engine and tender, weighing seventy tons. The time occupied was three hours, nine minutes. The trial proved satisfactory. An analysis of the peat by Mr. Ricard was given, which showed it to contain but a trace of sulphur, and no phosphorus, which rendered it peculiarly adaptable for iron-smelting and other purposes, where the presence of either of those bodies was so pernicious.

"Particulars were given of an experiment, on a practical scale, by Mr. G. Murrall, at the Creevelea Iron Works, Leitrim, in which condensed peat was used for smelting iron-ore. The iron was equal to any charcoal iron. It was tried by Mr. Anderson, C. E., whose report showed the strength of the iron to be forty per cent. above ordinary Scotch pig.

"The peat was also tried in a puddling furnace, at the Mersey Steel Works; and the iron produced therefrom was drawn into tubes and T irons, — a very high test.

"The question of producing gas was gone into, and experiments by Mr. Jones and Mr. Versmann, the former engineer and the latter consulting chemist to the Commercial Gas Co., were detailed, and tables of results given, which showed the superiority of condensed peat over coal in this particular. It was proved to yield a larger quantity of gas in a shorter time than coal.

. . . "The author concluded by observing that the question of fuel was a most important one, especially so in its economical bearing; and expressed a hope that the merits of peat, as a fuel, would be allowed to weigh, that the subject might be investigated with that vigor to which it was entitled."

The following is abridged from the "London Mechanics' Magazine": —

"The upper portions of a peat-bed abound with roots and coarse fibres, and produce an inferior fuel, which will not stand the blast nor make good charcoal. To make good fuel, the roots must be removed and the smaller fibres broken up. A machine does this, and makes 'condensed' peat, which burns freely, bears a strong blast, gives great heat, is smokeless, and leaves less ash than average coal or coke. Two and a half to three tons of peat make one ton of excellent charcoal. The heating power of condensed peat has been proved to be superior to that of coal; and it is well adapted to steam-engines, marine, stationary, or locomotive. It saves half the time of getting up steam, and will do double duty as compared with coal. The absence of smoke and clinkers, and the preservation of the grates and fire-boxes from the effects of sulphur, are important additional advantages. This peat has been tried on a river-steamer with success. The vessel was under steam 2h. 20m., during which time only 12 cwt. was burnt, the average consumption of coal being 12 cwt. *per hour;* and in this case the full effect was not obtained, as the grate-bars were too far apart for peat, and a portion fell through.

"The locomotive superintendents of three railways in Ireland made a trial of condensed peat on the Belfast and Northern Counties Railway, to test its fitness for locomotives. During a trip of twenty-seven miles, there was an excess of steam, though the fire-door was continually open, and the damper down, for the greater part of the distance. The pressure at starting was 100 pounds. The commencement of the trip was up an incline of 1 in 80, four miles long, with double curves. While ascending this incline, the pressure rose to 110 pounds, and afterwards to 120 pounds, with the fire-door open. The speed was forty miles per hour. While running, there was no smoke, and little at the stations. The fire-box was examined at the end of the trip, and very little clinker was found; and the smoke-box was free from cinders and dust, — a proof that the fuel had stood the blast well; and it is the recorded opinion of the experimenters that the peat was, in every respect, well suited for locomotives.

"In view of such facts, the wonder is that it has not come into more extensive use. One reason why it has not may be in the limited quantity made for steam purposes, the greater value of peat lying in its conversion into charcoal, for which it is used with the best results. Another cause may be the reluctance to leave the beaten track.

"This substitute for coal deserves attention in connection with the working of underground railways. It is agreed that a radical change must be made in the motive-power; that even Mr. Fowler's new engine must be superseded by one having no fire at all; and that the engine must contain water already heated to a temperature necessary to produce a sufficiency of steam to work the required distance without combustion in the tunnel. In this extremity, why not try the fuel which did so well on the Irish railway trip? The absence of smoke and sulphurous vapor should be a sufficient inducement, independent of the economy that is very probable."

A foreign correspondent, before referred to, writes, "For all the locomotives of the railroads in South Bavaria, peat is the only fuel;" and the economy effected by its use, in the wear and tear of the engines, is stated by the officials, in their reports, to be very considerable.

At a colliery in the north of England, connected by a railway with large iron-works, several trips were made with condensed peat fuel, a number of empty trucks being conveyed to the pit's mouth, and brought back loaded. The fuel served admirably, and all present were satisfied it did quite as much duty as coal.

On one trip, however, they were obliged to wait some time at the pit's mouth, the load not being ready. The fireman, from some cause, neglected his fire; and consequently, when the trucks were loaded, and the train ready to start, he found the steam-gauge indicated a pressure of only sixty pounds, and the fire very low. This alarmed him, as the return journey was up a steep incline, and he had already been delayed very considerably. However, in *three minutes* after throwing into the furnace a few shovelfuls of peat, the steam rose to one hundred and ten pounds, and the train was soon speeding its way back. Had coal been thrown on, instead of peat, it would have smothered the fire, so little was there in the furnace at the time.

The above illustrates very clearly the freedom of combustion which characterizes peat, and the intensity of heat produced by it.

At the Horwich Works, the fuel was tested against coal under the boiler there. This was done on two consecutive days, the fire having, on each occasion, been raked out the night previous. The following results were obtained: Coal got up steam to 10 lbs. pressure in 2 hours 25 minutes, and to 25 lbs. pressure in 3 hours; peat-fuel got up steam to 10 lbs. in 1 hour 10 minutes, and to 25 lbs. in 1 hour 32 minutes; 21 cwt. of coal maintained steam at 30 lbs. pressure for 9⅓ hours; 11¼ cwt. of peat-fuel maintained steam at the same pressure for 8 hours. But, in addition to this a large economy is effected by the use of peat-fuel for the generation of steam, in the saving of boilers and fire-bars from the destruction caused by the sulphur in coal, from which peat is free.

We have seen a statement of a locomotive running upwards of three months, over seventy miles of road, and using *peat*, which show a saving of more than thirty per cent. by weight over coal, using coal furnaces and flues, with dampers down and fire-doors open all the time.

A trial on the Paris and Lyons (France) Railroad is thus reported by the engineer: "We got up steam with peat in thirty minutes, coal requiring two hours. We ran sixteen miles to a gravel pit, up a steep grade, and from there took a load of 136 tons, eighty miles farther, when the blaze escaped a considerable distance above the chimney, which became red hot, and the boiler covering taking fire, we had to stop and extinguish it. After repairs we returned to Paris at a speed of thirty-eight miles per hour, the heat again increasing as we advanced.

"The fuel having no smoke and much gas, keeps up a constant hot flame. The pressed peat gives far better results than that which is not pressed. In fact, while using it, the generation of steam was so rapid that I stood with my hand on the valve lever all the time, fearing an explosion."

Although numerous experiments and trial trips have been made in this country with peat for fuel, it has not yet been used to any great extent upon any of our principal lines, owing mainly to the fact that it has not been manufactured in any one place on so large a scale as to afford a constant supply, indeed, notwithstanding it has been produced in numerous places, few engaged in the enterprise seem to comprehend the enormous amount of this or any other kind of fuel which is requisite to supply the market. The arrangements for the present year, in some places, however, bid fair to be on a scale somewhat commensurate with the fuel interests which are likely to draw on them for supply.

It is well known, that in every experiment tried upon railroads, the testimony of engineers and practical men has been unanimous in favor of peat, even when it has been used in its crude state, and under every disadvantage.

Although many cases of this kind have come to our knowledge, from sources which we cannot but consider reliable, still we have experienced great difficulty in obtaining written reports covering that amount of detail as to time, weight, service, and comparative value which we should be glad to make record of.

We select a few, which give the general character of the whole, and from which it cannot but be apparent that the fact is clearly demonstrated that peat is *the* fuel for railroad service.

Some of its advantages are plainly discernible, and may, perhaps, be briefly stated as follows : —

It ignites readily and burns freely, generally with a large volume of flame. Combustion appears to be almost perfect, with a very clear and intense heat, producing no cinders, no sparks, no soot, very little smoke, and no clinker; the consequence of which is, that under a boiler, steam is generated very much more quickly than by coal, the flues and tubes of the boiler are kept free from soot, clean and bright, and therefore in better condition to make the heat available, and the grate-bars are not burned out and injured as with coal, while on the score of comfort to travellers, it may be said that the annoyance and actual suffering occasioned by cinders, sparks, and smoke, which, in spite of the numerous devices for consuming them, we are now constantly subjected to, are, by the use of this fuel, entirely obviated; and so clearly are these advantages demonstrated, that we are satisfied that wherever this fuel is fairly tried, every railroad will endeavor to have its own peat bog and fuel manufactory.

Another writer adds, "The economy of peat in the matter of burning out grates and furnaces is well worth the attention of railroad men and manufacturers. The average destruction of locomotive furnaces on the New York and New Haven Railroad is said to be more than two per day. Sometimes a furnace is ruined by a single trip of the engine. Peat will destroy no more furnaces than wood."

N. F. Potter, Esq., of Providence, R. I., President of the Narraganset Brick Company, informs us that he has used peat under their large boilers for several months, with highly satisfactory results as to its heating properties, and at a large reduction of expense as compared with wood or coal.

He has also found it equally serviceable and economical in running a small steamboat, likewise for a portable engine used for occasional out-door service.

A trial was made on the New York Central Railroad, Jan. 3, 1866, of which the following account is furnished us by the master machinist: —

"Engine No. 248, built at Schenectady Locomotive Works — cylinder 16-inch bore, 24-inch stroke, 5 feet driving-wheels: Michael Cosgrove, engineer, Michael Fox, fireman. Left Syracuse at 8 o'clock and 40 minutes (40 minutes behind time), with 25 empty 8-wheel box freight cars. Started with 120 pounds of steam: the engine worked well, and took us along pretty sharp, as we made up the 40 minutes in going 25 miles, and arrived at Port Byron on time. The steam did not run below 120 pounds any of the time, and was often from 125 pounds to 130 pounds. When the engine was working the strongest she would steam the best.

"We made time all the way very easy, although we had a strong head wind all the way, and snowing at times quite fast, and very cold. We took on a trifle over four tons of peat at Syracuse, which was all we had. We could have run to Fairport with it (71 miles) if we had not been detained at Palmyra about one hour and a half, waiting for a break-down. We had to keep our fire up while waiting, and used peat enough to take us much farther. We took wood at Palmyra. After leaving Palmyra we used our peat up before reaching Macedon, and finished the trip with wood. Five tons of peat would have taken us to Rochester with the train we had, although it was a very bad day. The same engine would have used about 3½ cords of wood running to Palmyra, while we used not quite 4 tons of peat for the same distance. It gave us as much steam as wood, and burned a beautiful fire. Our trip was a perfect success, and I am sorry that there was not more present to witness it. We used a coal-burning grate.

"I am confident that we can use peat in locomotives for fuel, as well as for stationary engines, with the peat properly cured, and the right kind of grate used for burning it in."

The "Hartford Times," of June 5, 1866, furnishes the following account of two trips made on the Hartford and Springfield Railroad: —

"Some facts were developed on the experimental railroad trip from this city to Springfield, last Saturday, with peat for fuel on the locomotive, which are worthy of the public notice, in view of their bearing on the subject of our undeveloped natural wealth, and important questions of economy.

"Without attempting full speed, the trip to Springfield was made in 40 minutes, including a stop of three minutes at Thompsonville, which is five minutes better time than the regular express train makes.

"On the round trip, 1400 pounds of peat were used (the box contains between two and three tons; so it will be seen that peat is by no means too bulky for long railroad distances between cities). It is thus proved (and it corroborates the results of the previous trial) that, compared with coal, a trifle over one ton and a quarter of peat is equivalent, on a locomotive, to a ton of coal.

"This ton of coal, however, costs the railroad folks, say seven to eight dollars.

"Peat, on the contrary, can be furnished in abundance at $1.50 per ton! The Hartford and New Haven Railroad Company have paid out during the past year over $100,000 for coal, using, perhaps, 12,000 tons.

"They now own a peat-bed in Berlin, contiguous to the line of the railroad, and containing eighty acres of peat, of great depth, from which they can supply the machine shops and locomotives with peat, at a saving of more than seven tenths of the annual cost of coal.

"There was found to be but very little smoke from the peat; what there was, was light, thin, and not offensive; not one tenth of the smoke produced by wood, and, what is still more important to the travelling public, no cinders. The whole substance burns to ashes. It was found to burn the freest, and to give the greatest rate of speed, when the furnace full had burned away to about one third full.

"It generated steam faster than either wood or coal, gaining so rapidly that the furnace had to be thrown open; and the last six miles of the return trip was made without using a particle of fuel.

"The capacity of peat to generate and maintain a high and equal rate of steam power was tested a few days since by the trial of this fuel on a train of ten cars, very heavily laden with Portland stone. Before reaching this city the brakes were put down so as to increase the resistance to an amount equal to five more cars, without very sensibly affecting the motion."

A trial of peat-fuel was recently made on one of the railroads leading from Boston, particulars of which we have taken pains to ascertain from the superintendent of the road, although he states that the fuel was used, not for the purpose of making an accurate test at that time, but simply to ascertain something of its burning qualities, and the arrangement of fire-box, &c., necessary for making, at a subsequent time, careful test trial of its value for locomotive use, compared with wood, which is the fuel now used on the road.

The section over which the trains were run is a branch road of only seven miles, — down and return; of this seven miles, four miles average a grade of 45 feet, and 2½ miles average 65 feet to the mile, while one mile of the 2½ is at a grade of 80 feet.

The quantity used was not weighed. The engine was a wood burner, with large fire-box and very strong exhaust, kept for occasional extra service, and noted as an enormous consumer of wood.

Neither the superintendent nor the engineer had ever used this kind of fuel, and neither had ever seen it used. The trips were made, down and up, on good time, and with steam at 100, and blowing off at that, except while going up grade, when, at the hardest point, it fell to 70, but rose to 100 readily before the end of the trip. On one trip ten pounds of steam was gained while going up the 80 feet grade.

The superintendent says, "We made good steam, and run our trains on time, and that is what was never done with coal, on the first time trying, by a long shot. It makes a splendid fire, gives an intense heat, with no cinders, no soot, and very little ashes; and I see no reason why it should not be used on locomotives, if they are only properly fitted for it. We want pure peat, free from sand, in order to avoid the clinker which melted sand will make, and then all we want is to learn *how* to use it; but that we can learn much easier than we learned how to burn coal. I am going to try it again; but I think I have done well, certainly for the first time, the whole thing being entirely new to us, and considering also that we tried it on an old engine which we never use except in case of emergency, because she burns three cords of wood where the other burns only two."

We also learn of trials made on the Hudson River Railroad, the Eastern, and the Vermont Central, each of them with uniformly successful results.

The following reports of the use of this fuel on the Grand Trunk Railroad are conclusive as to its merits, and will be found to give details and suggestions which cannot fail to be of importance to those who would investigate the matter with a view to practical operations: —

An experiment was made with well-dried peat fuel upon engine No. 158, five-feet driving wheels, sixteen-inch cylinders, and twenty-six-inch stroke, drawing twelve loaded cars.

Distance run per ton of 2240 lbs. of fuel . .	40.33 miles.
Fuel used per mile	55.54 lbs.
Greatest pressure of steam	140 "
Least pressure of steam	100 "

During the experiment fuel was put on in small quantities, no smoke issued from the stack, a steady, brilliant, white fire was kept up, and steam generated with great rapidity. The damper was kept closed, and air admitted through a slot in the furnace door. Not an atom of ash or cinder was left in the smoke-box, ash-pan, or upon the wire gauze of spark-catcher. The grate inside was one of Lester's patent, having a well in centre, with horizontal openings to admit draft. The bottom of fire-box was scarcely ever entirely covered with the fuel, the steam being generated too rapidly to allow of a large quantity of fuel being put into the furnace.

Experiment with green peat fuel, containing twenty-five per cent. of water, upon engine No. 65, six-feet driving wheels, fifteen-inch cylinders, twenty-one-inch stroke, drawing an express train of *nine* passenger cars, all *heavily* laden, from Montreal, going west, Oct. 3, 1866:—

Distance run	101 miles.
Fuel used	8000 lbs.
Fuel used per mile	79 "
Average speed, including stoppages, per hour .	23 miles.
Greatest pressure of steam	123 lbs.
Least pressure of steam	90 "

This experiment was one to show whether, with an engine out of order and very much overburdened, steam could be made with green peat in sufficient quantities to meet an unceasing demand during the whole time of running.

The grate was one of Mr. Eaton's patent, with horizontal openings, and the furnace door had a slot to admit air.

In ascending grades the pressure of the steam gauge invariably increased. The damper was nearly closed at all times, the slot in the door open nine by two inches, and with never more than from six to nine inches of fuel in the fire-box. Abundance of steam was raised; and for a distance of many miles the pressure of steam did not vary.

On the return trip next day, with a similar weight of fuel, and train of six passenger cars, —

The total distance run was	112 miles.
Fuel used per mile	71 lbs.
Average speed, including stoppages	22 miles.
Greatest pressure of steam	125 lbs.
Least pressure of steam	85 "

Experiment with engine No. 65, in good working order, and with peat fuel containing about 20 per cent. of water (express train, consisting of six passenger cars): —

Total distance run	177 miles.
Total consumption of fuel	7936 lbs.
Consumption per mile	45 "
Maximum consumption between stations . .	60 "
Minimum consumption between stations . .	30 "
Average speed, including stoppages	25½ miles.
Greatest pressure of steam	125 lbs.
Least pressure of steam	86 "
Distance run per ton of fuel	50½ miles.
Cost of fuel for the trip, at $3.50 per ton .	$12.25
Cost per mile run for fuel	7 cents.

During the experiment the damper was kept partly open, and the slot in furnace door of an area about nine inches by two inches. Fuel was fed in sufficient quantities to hide the bottom of the fire-box, and throughout the trip there was not the slightest deficiency of steam.

On the return trip the consumption of fuel was less, the train being lighter.

How these results compare with those obtained from coal and wood, will be seen from the following statement, recently published, derived from experiments made on the Boston and Worcester Railway in August last: —

Average miles run to one ton of coal	59.91
Average miles run to one cord of wood	40.09
Average miles run with one ton of peat, as above .	50.50

The value of these results will be better appreciated from a comparison of cost of the several fuels.

For the amount of coal, wood, and peat required to run a locomotive the distance mentioned in the foregoing experiment, say 177 miles, the cost is stated as follows: —

Coal, 2.95 tons, at $10	$29.50
Wood, 4.41 cords, at $7	30.87
Peat, 3½ tons, at even $5	17.50

The country at large, as well as the railroad interest, is concerned in these developments in reference to an article of fuel which has lately attracted so much attention.

Another experiment was made on the same road, with the view of determining whether, with a diminution of the blast, the same quantity of steam could be generated as obtained on former occasions, with the blast usually employed for wood.

For this purpose the nozzle of the blast-pipe was increased one half of an inch, or from 2½ to 2¾ inches diameter, making an additional area of fifty per cent. The same engine, No. 65, was employed as on former occasion. Mr. Eaton accompanied the train, for the purpose of putting the fuel to the severest test possible.

The engine, when strained to the utmost of its power in ascending heavy grades, or in quick running on a level road, produced abundance of steam, and kept blowing off the whole time. By this diminution of the blast additional power was gained, and the consumption of fuel smaller than on any previous occasion.

Still another experiment made on the same road in November, 1866, affords in detail, the results effected with enlarged blast-nozzles of the engine. Engine No. 158, with 16 freight, 1 passenger, and 1 baggage car, run from Montreal to Prescott Junction, 112 miles.

Total rise in grade	260 feet.
Total weight of 18 cars and freight . .	665,000 lbs.
Distance run	112 miles.
Lost time made up in running between Vaudreuil and Matilda, 75 miles	110 minutes.
Total weight of peat consumed	7450 lbs.
Value of fuel, at $3.50 per ton	$11.65
Fuel consumed per mile run	66½ lbs.
Cost of fuel per mile	10 cents.
Number of car-miles run	2016 miles.
Fuel consumed per car-mile run	3.69 lbs.

Cost of drawing a car, containing over ten tons of freight, one mile, a little over half a cent.

The engine was in the same condition as for burning wood, with the exception of the blast-nozzles, which were enlarged from 2⅜ inches to 2¾ inches diameter, or 34 per cent.

The engine driver had never before seen peat burn. The fuel used on all these trials was from the Bulstrode Peat Works, owned by James Hodges, Esq., of Montreal.

Messrs. Parsons & Co., refiners of oil, at East Boston, required for their process super-heated steam to degree and extent which it was found extremely difficult to obtain with wood or coal of any kind; and under the most favorable circumstances required two furnaces, one for generating and the other for super-heating. A faithful trial of peat was made, in view of statements concerning the *intensity* of heat generated by it, detailed report of which was promised us, but has not yet come to hand. The result, however, may be briefly stated as follows: The peat fuel proved to be entirely satisfactory. Steam was not only generated, but super-heated to the full extent and temperature required; and it was satisfactorily demonstrated that with this fuel the whole service demanded might be had from one furnace instead of two, — an item of no light importance. The fuel was from the works of the Boston Peat Company.

A trial of the same fuel was recently made at the Lowell Bleachery, under their large boilers (engine, 500 horse power), of which we have only verbal, though *reliable* report. It made more steam than coal, and accomplished with ease the severest service required at the works. The superintendent says he considers it a perfect success, and that all now required is to adapt the fire-boxes for it, and *learn how to use it.*

This latter remark will doubtless be found to express the opinions of numerous parties when commencing to use peat fuel. That it is a *good* fuel is universally conceded, which is much more than was even admitted for coal when that was first introduced. The important questions now are as to the arrangements of furnaces, grates, drafts, etc., and the manner of using the fuel to the best advantage; in other words, "the people" have to *learn how to use it*, but the task is an easy one.

Peat for Domestic Purposes.

We have heretofore dwelt mainly upon the use of peat in the manufacture of iron, and for steam purposes as relates to manufacturing and transportation interests.

For domestic purposes generally, we can have no doubt, from our own experience and the testimony of others, that ton for ton, it is of equal value with, and in some respects more economical, than coal.

It is not our habit to commend this fuel as invariably and for all purposes superior to all others, but in a multitude of cases where parties have had small lots for trial, they have brought their ready testimony, that it was the best fuel they ever used; and this, we firmly believe, is to be the general verdict.

In saying this, however, we wish it to be borne in mind distinctly, that with peat fuel, as with every other kind of fuel, its value for service depends as much or more upon the manner in which it is *used*, and the methods and appliances for using it, as upon the peculiar qualities and characteristics of the fuel itself.

It can be burned in open grates, close stoves, furnaces, ranges, and all the ordinary variety of heating apparatus in use in dwelling-houses. As regards open grates, ranges, and furnaces, it will be found that the fire-pot, or receptacle for fuel, should be smaller in area and of less depth than are at present in use for coal. It requires to be renewed somewhat more frequently than coal, and should be burned with very much less draft, indeed so soon as the fuel is once well ignited, it will generally be found desirable to close the draft almost entirely.

On subsequent pages, under the head of "How to use Peat Fuel," we shall offer some suggestions, which are of importance in this connection.

We have taken no pains to obtain written certificates bearing upon its use for these purposes, and probably sufficient testimony has been given, incidentally, on preceding pages; but we have at hand letters from two gentlemen, for many years residents of Lexington, and well known there, which are sent to us voluntarily, and which, as exceptions to the general rule, that "a prophet is not without honor save in his own country," are perhaps to be considered as even stronger evidence than testimony obtained farther from home, and are illustrative of what we have learned to be the experience of many others.

Mr. S. W. Robinson, of Lexington, writes, —

"I have of late been using peat, which was prepared at your works, and am favorably impressed with its importance as an article of fuel. That it is *better* than the coal with which families in this vicinity have been supplied for two or three years past, there can be no reasonable doubt. It is easily ignited, burns freely, and leaves neither cinders or other impurities behind. One great advantage it possesses is, that it may be used in smaller quantities than coal, and may be burned anywhere, in grate, stove, or hearth. I am persuaded that it is well adapted to culinary purposes, and is not only *better*, but more economical than coal or wood, which is surely an important desideratum."

Captain Wm. D. Phelps, of Lexington, writes, —

"We used two tons of your peat during the summer and autumn, in a Stewart's cooking stove of medium size; and for all purposes of the kitchen we give it the preference to wood or coal, especially for baking and broiling meats, &c., and for heating irons; and, as a peat fire can be graduated to suit the weather and the occasion, it is far preferable to coal. It makes a hotter fire also, and takes but fifteen minutes from the starting of the fire to heating the oven ready for baking. My neighbors who have used your fuel are satisfied and pleased with it."

The following extracts from an article in the "Boston Transcript," are also to the point: —

"Possibly a few words in relation to *peat* may not be without interest to your readers.

"I have used it for the past two months, both at my residence and place of business, in furnace, range, and open grate; and am convinced that any who try it will be surprised, as I have been, to find how rich an article of fuel we have lying in immense quantities in the swamps about us, and capable of being produced at moderate cost. I have experimented with it in various ways, and am decidedly of opinion that, although the process of manufacture is doubtless open to considerable improvement, the article, as at present furnished, is, in many respects, superior to coal. It ignites readily, burns freely, leaves a handsome white ash, but no clinker, gives an intense heat, and is easily managed. For open fires it is charming.

"Many years ago, as doubtless many of our oldest citizens will recollect, peat in its crude, unmanufactured state, was sold in Boston in considerable quantities. It is extensively used in many places in New England, and that, too, in preference to wood, even where wood is abundant. The high price of fuel the past season has led many to seek for it; and I have heard of those who have laid in considerable quantities of it in Newton, Reading, Lynn, and in numerous places on the Cape.

"In many parts of Europe it constitutes the principal article of fuel for domestic purposes, and is used to some extent on locomotive engines and in large manufacturing establishments. For the manufacture of iron and steel, it is said to be superior to any other fuel in existence. Is not the fact that abundant stores of this fuel are 'on deposit' in various sections of New England worthy the attention of enterprising men?"

It is quite probable that peat will soon be manufactured in a neat form, specially adapted for kindling purposes; and movements are also projected for introducing it for sale in our cities, in boxes, in much the same manner as kindling-wood is now sold, while for ordinary purposes it will be sold in bulk, like coal.

Intensity of Heat generated by Peat.

It is an acknowledged fact, that peat produces an *intense* heat, — a feature of so much importance as to entitle it to prominent mention and careful consideration. Its virtue in this respect is much increased when properly prepared, solidified, and dried, and it reaches its maximum of heating power when solidified and charred or coked. Mention has been made of its peculiar qualities in this respect, in several of the statements on preceding pages, but their importance will be more clearly comprehended when taken in connection with the facts in an interesting article on the "*Calorific Value of Fuel*," which we find in the "American Railway Times," and which contains remarks so pertinent on this point, that we quote as follows: —

"There are, in all, five important kinds of fuel only: these are wood, peat, coal, charcoal, and coke; the first three being natural, and the last two artificial fuels. The elements of which each of these is composed are practically identical, the differences of character being due to the proportion of those elements entering into the composition of each kind of fuel; and according to those proportions, each fuel takes its relative position in the scale of value. Taking the comparative chemical composition of the various kinds of fuel, according to Dr. Machaltie, their percentage stands thus:—

	Carb.	Hyd.	Oxyg.	Nitr.	Sulp.	Ash.
Wood (dried at 280° F.) .	50.0	6.0	42.9	1.0	. .	1.0
Peat (dried at 220° F.) .	57.0	5.5	31.0	1.5	. .	5.0
Coal	85.0	5.0	4.0	1.0	1.0	4.0
Charcoal	87.0	3.0	7.0	. .	. .	3.0
Coke	92.0	. .	. .	. .	1.5	6.6

The *amount* of heat produced by fuels in their combustion does not always constitute their relative value. For some purposes, it is apparent that this would be the best criterion, but as a rule, in metallurgic processes, the *quantity* of heat is of far less importance than the *intensity*, or power to raise substances to the highest temperature; and the fuel which affords the greatest quantity of heat is sometimes incapable of producing the greatest intensity.

"In determining the intensity of the heat produced, it is necessary to know the available quantity of heat produced in the combustion of a pound of fuel, the weight of the products of combustion, and the quantity or number of units of heat required to raise the products of the combustion of a pound of fuel one degree Fahrenheit.

"Where very high temperatures are required, the fuel which should be selected ought to approach as near as possible to pure carbon in its composition, and for the reason that carbon is the best substance for the purpose.

"We now see the reasons for making coal into coke, and wood into charcoal. Coal cannot produce a temperature equal to that obtained from coke, neither can the temperature of wood be compared with that of charcoal. And this results from the relative accession of carbon, and reduction of oxygen and hydrogen in them. This must be referred to the great difference between *quantity* and *intensity* of heat. If we cannot raise sufficient steam from a boiler by the use of one ton of coal, we can easily meet the point by burning two tons; but, if the fusing point of metal cannot be attained with one ton of coal, it by no means follows that any additional amount of fuel will insure the required result. The great distinction to be observed is between *quantity* and *intensity* of heat. The first of these two conditions depends upon the quantity of fuel; but the last is referred entirely to the quality of fuel.

"Twenty tons of coal will not give a temperature so great as that afforded by one ton of coke."

The intense heat generated by *peat-fuel* is a subject of frequent remark, and will eventually be dwelt upon, we think, as a very important consideration in estimating its value.

How to use Peat.

It should be borne in mind that peats differ in quality and characteristics, and consequently in their heating properties and value, full as much as wood and coal in all their varieties; and that the results obtained from peat-fuel, as from any other, will, in all cases, depend very much upon the manner in which it is burned.

We are told that a certain service is obtained from a boiler with one cord of wood or one ton of coal, but unless we know whether the wood be pine or hickory, or the coal Pictou or anthracite, we have no certain data from which to make accurate or even approximate calculations; so also of peat, it is necessary to know somewhat of its characteristics in each case reported, in order to form a correct estimate of its relative value for the service done. The terms by which to designate these characteristics and qualities, so as to be generally understood and adopted, have yet to be determined upon, but the necessities of the case will doubtless bring them out in good time.

That peat is a *good* fuel is universally conceded, which is much more than was even admitted for coal when that was first introduced. The important questions now are, as to the arrangements of furnaces, grates, drafts, etc., and the manner of using the fuel to the best advantage; in other words, "the people" have to learn *how to use it*, but the task is an easy one.

It is an acknowledged fact that very few people have any correct idea of the *economical* use of fuel of any kind.

Everybody is supposed to know how to burn wood, but very few burn it economically, or even prudently. The waste of wood is immense.

So also of coal. It was a long time after hard coal was introduced before the community became even tolerably familiar with its use; but at the present time it is otherwise. With plenty of kindling wood, an abundant supply of coal, a clear grate and a strong draft, a good fire is started and kept up, "regardless of expense;" but for a moderate fire on a mild day, or for light service, few are skilful enough to kindle or maintain it.

Such being the case in regard to wood and coal, is it reasonable to expect that any one, at the present time, is *thoroughly* posted as to the *best* and most economical methods of using peat?

Our own experience has taught us that, as a general thing, peat-fuel should be burned in smaller area and bulk than coal, but renewed in small quantities somewhat more frequently, and with very much less draft. The heat is clear and intense. The quantity of kindling wood required is very small.

It is a matter of no slight importance to ascertain and introduce the best stoves, furnaces, grates, ranges, &c., for burning peat.

Thus far little has been done in this line, though a few enterprising stove dealers have already turned their attention to it, with good prospect of success; and their advertisements are beginning to announce "*Peat* Stoves," &c.

The appliances required are simple, and there will be little difficulty in producing what is wanted, or in altering and adapting, by very simple changes, many of the favorite styles now in use for coal.

It is not, however, in the matter of stoves and appliances for domestic purposes alone, that these improvements are called for, but in fire-boxes for locomotives, furnaces under boilers, and wherever fires are to be used in the arts, manufactures, &c.; and it is clearly to be seen that this opens a wide field for experiment and invention among boiler and engine-builders.

As relates to the use of peat-fuel for locomotives, Mr. Hodges, before referred to, has offered some remarks which correspond so nearly with our own observations and the general tenor of statements made to us by others who have investigated the matter, and are, withal, so clearly expressed, that we quote: —

"Peat-fuel, even with the limited experiments as to the

best mode of using it economically, has proved itself equal if not superior to any known fuel, and it is no more than reasonable to anticipate greater results when its use becomes general, and furnaces are expressly adapted to its use.

"As locomotives are now constructed for the combustion of wood, coal, or coke, the waste of steam power to create a blast or draft is enormous, it being estimated by competent authority that two fifths of the whole quantity of fuel consumed is expended for that purpose. Now, well-dried peat requires but very little draft through the furnace bars, it being necessary, for a perfect combustion of the immense quantity of gas that it gives out, to admit air through the furnace door. It is therefore much more than probable that, by altering the blast to meet the limited requirements of peat-fuel, at least twenty per cent. additional power will be given to all peat-burning engines, or a corresponding decrease in the quantity of fuel used may be effected.

"No sparks issue from the smoke-stack of a locomotive when burning peat-fuel, even with the present enormous blast; and when they are especially adapted for it, fires from sparks will be unknown.

"Peat-fuel, containing from twenty-five to thirty per cent. of water, may be burned in a locomotive with a blast, and arrangement of fire-box precisely the same as for wood, and used in the same way, with this difference only, that with wood the fire-box is always kept full to the top, while with damp peat not more than six inches covering of the grate is necessary. In ascending long inclines, or with an overloaded engine, it may be necessary, perhaps, to increase the quantity to nine inches; but under no circumstances has the writer ever seen a twelve-inch covering to the fire-bars requisite.

"When it is considered that in burning a ton of green peat, containing only fifteen per cent. of moisture in excess of ordinary air-dried peat, thirty-three gallons of water have to be dried out of it or evaporated during the combustion; and, in addition, that the weight of solid matter in the ton of fuel is reduced fifteen per cent. by the water it contains, the difference of work performed by dry peat over that of wet is not so great as might be expected. This, however, may be accounted for by the little experience we have hitherto had with the fuel, and also from the fact that locomotives have not been adapted to its use.

"The amount of blast required for green peat is not so great as that required for wood; but it burns well in a furnace arranged for consuming wood.

"For dry peat, very little blast is required; and when burning in engines adapted for wood, the fuel has to be applied in such small quantities that it is scarcely possible to keep the fire-bars covered without raising more steam than is required. It therefore seems apparent that, although an approximate maximum of work may have been got out of green peat, the experiments with dry fuel need repeating many times to give any adequate idea of what it will do in properly constructed furnaces, and with a suitable amount of blast."

The strong exhaust in most of the locomotives at present in use is by no means needful or favorable for obtaining from this fuel its best service; and there can be no doubt that in this respect a radical change will be found desirable as one of the essential features of a good peat burner. The area and depth of the fire-box, and the arrangement of the grate-bars, for both locomotive and stationary engines, are also likely to be modified to a considerable extent.

"Chemical analysis shows that peat, weight for weight, contains only three fifths of the heating properties of coal, and it is therefore the opinion of many that it is little more than half as valuable for raising steam.

"Now this is all very well in the closet; but as practice shows that, even with the best constructed furnace thirteen per cent. only of the heat-giving properties of coal are utilized, there is still a pretty good margin for peat, and a possibility that, by being able to economize a greater percentage of the heat-giving properties it contains, to make it do double the work of coal.

"It is true that Dr. Whelpley, of Boston, by his wonderful pulverizer, reduces refuse coal to an impalpable powder, and, by a peculiar mode of combustion, makes this dust do six times the work of the best coal, at a cost perfectly insignificant compared with the result. That he will succeed eventually in making a revolution in the use of fuel there can be no question; but as peat-fuel can be reduced to dust as easily as coal, there is still a hope that peat-fuel may keep its position as the best fuel in the world."

We cannot claim to be sufficiently well informed upon this very important part of our subject to undertake to give definite instructions applicable to all cases. Many useful hints bearing upon it have been given on preceding pages, and it must now await the result of experiment and investigation, which, with the "inventive genius and combining skill" already enlisted in this great enterprise, are sure to meet the requirements of the case. The field is a wide one.

The following, which we cut from the "American Gas Light Journal," is pertinent in this connection, as relating to economy in the use of fuel.

"Take as a period the last fifty years, and see what improvements have been made in this direction. While our forefathers were content to warm the humble cottage by the aid of the 'fire-place,' which occupied one side of the dwelling, whose capacious jambs required the immense 'back-log' and 'fore-stick,' and the various components to form the huge pile for a respectable fire, their children employed the 'box' and 'Franklin' stoves, &c., which used both coal and wood, and were, scientifically, an improvement in the degree of radiation attained, and because they did not carry the most of the caloric up the chimney in the tempest.

"From this we come to the more modern and scientific appliances for heating, cooking, &c., embracing heaters and registers, radiators, base-burners, smoke and gas-consumers, patent cooking-stoves, ranges, galleys, and numberless other inventions. So, also, in equal or greater degree, has there been improvements in the various processes of smelting, and in steam engines, both land and marine, by the aid of improved draft or blast, by return, horizontal, and inclined flues, patent jackets, grates, condensers, &c., to more completely consume the smoke and gases, to increase the radiating surfaces, and various other improvements, until, when we look back to the old methods, we smile at their primitiveness and inefficiency.

"Science is progressive; and the inquiring mind will ever be on the alert to improve the various appliances now in use, where coal or other fuel is employed, to more completely utilize and economize the caloric evolved during combustion, to devise more economy in mining, and to discover and apply substitutes for coal fuel.

"It is safe to say that by modern improvements we get from one to two hundred per cent. more caloric, or work, — which is but another name for it, — from a given amount of fuel consumed, than we did fifty, or even twenty-five years ago; and there is no reason to believe the inventive genius of the age will not improve upon the present methods of consuming fuel economically, during the next hundred years, in a similar ratio at least."

Gas from Peat.

For *gas*, its properties have been tested in several places in this country, as also in Paris, Plymouth, Dartmoor Prison, Dublin, Munich, Kempton, and other places in Europe, with uniformly satisfactory results, but varying considerably, according to the character of the crude material. Its yield may be said to be about the same as the Newcastle coal; but, in most cases, the light produced has been pronounced superior in brilliancy and power.

Gas of an excellent quality for lighting, and in large quantities, can be produced from some kinds of peat; and in the ordinary progress of events must, we think, be extensively used, and that at no distant day. Numerous and very successful experiments in this direction have been made, and results published from time to time. As long ago as 1683, J. J. Beecher published an account of his having produced gas from common coal in England, and from peat in Holland; and, from that time to the present, abundant proof of its value and easy production has accumulated.

For purposes of making and refining iron, the gas

produced from peat has been extensively used in France, Germany, Prussia, and Sweden. In many places in Europe it is used for both purposes, and its consumption is regularly augmenting.

On Dartmoor, the peat is cut by the convicts from the prison, working in gangs; and, after being dried, it is carefully stored in one of the old prisons. From this peat, by a most simple process, gas is made with which the prisons at Princetown are lighted.

The illuminating power of this gas is very high. The charcoal left after the separation of the gas is used in the same establishment for fuel and for sanitary purposes, and the ashes eventually go to improve the cultivated lands of that bleak region.

Attempts were made here many years since to distil the peat for naphtha, paraffine, &c.; but the experiments not proving successful, the establishment was abandoned.

An article in "Silliman's Journal," 1855, after remarking upon discussions had in the city of Paris in connection with the renewal of the engagements of the city with the gas companies, goes on to say, "Attention has been called to the gas manufactured from peat, which for some time has been used in Paris.

"M. Foucault has been charged with measuring the comparative illuminating power of coal and peat gas, and the result is in favor of that of peat, its power being three hundred and forty-two, while that of coal gas is one hundred. The manufacture of peat gas is also more simple than coal. The peat, if put into an iron retort, heated to a low red heat, affords immediately a mixture of permanent gases and vapors, which condense into an oleaginous liquid, which two products separate on cooling. The oil is collected in a special vessel, and the gas passes into a gasometer. This carburetted hydrogen is wholly unfit for illumination, as it gives a very small flame, nearly like that from brandy. The oil from the peat is a viscous, blackish liquid, of a strong odor. It is subject to a new distillation, and resolved wholly into a permanent gas and hydrogen very richly carburetted. This mixture is strongly illuminating, giving a flame six or eight times brighter than the first, and of a more lively brilliancy. The two are mixed, and a gas of intermediate character obtained, which is delivered over for consumption.

"M. Foucault made his trials with a photometric method not then made public. Its unit was not a single wax candle, but a collection of seven candles, arranged in a hexagonal manner, with spaces of one centimetre. A single candle is liable to too much variation, a compensation for which is secured when a number are employed. By this method a mean of five determinations gave for a burner of peat gas a light equivalent to twenty-three and one fourth candles, and the same burner, with coal gas, six and three tenths candles.

"The illuminating power of the pure oil from peat, the illuminating material, *par excellence*, has been found at equal pressures, seven hundred and five, the intensity of coal gas being one hundred; and, with equal volumes, their numbers are as seven hundred and fifty-six to one hundred."

From the absence of sulphur in peat, the purification of this gas is much more easily accomplished than that from coal.

Mr. Paul, to whom reference has before been made, stated before the Society of Arts, in 1862, that for a long time he had manufactured gas from peat to light his works; that the bituminous black peat of Scotland produced gas of good illuminating power, which required no purifying process for ordinary purposes; and he considered the applicability of peat for gas-making a matter well worthy of consideration.

Mr. Keats, before the same society, said he had made experiments of a like character, and had obtained as much as ten thousand cubic feet of gas from a ton of peat; but the percentage of carbonic acid was so great, in some cases amounting to twenty-two per cent., that it operated as a prohibition of the use of the gas until some cheap means for removing that difficulty should be found.

Mr. Brunton stated that trials made with his prepared peat had shown not only that there was a large amount of gas in the peat, but that the gas was of such high illuminating power as to bear advantageous comparison with gas distilled from coal, while the difficulty of the carbonic acid gas was done away with, in a great measure, by the partial charring or baking of the peat before it was used for the manufacture of gas.

Mr. Versmann had conducted experiments, and came to the conclusion that this prepared peat was a most valuable material for gas purposes, and that it would produce from twelve thousand to fourteen thousand cubic feet of gas per ton, of an illuminating power exceeding that of ordinary coal gas, the amount of carbonic acid not exceeding ten per cent.; and although that was somewhat in excess of the average of coal gas, yet there were advantages in peat which more than counterbalanced the disadvantages arising from the excess of carbonic acid.

Professor Emmons, in his report on the Geology of New York, remarks upon peat, and says, "It contains a gaseous matter, equal in illuminating power to oil or coal gas." And again: "Peat furnishes an abundance of carburetted hydrogen, and hence may be employed for producing gas-light.

"Dr. Lewis Feuchtwanger, of New York, has made known to the American public the experiments of Merle, a director of a gas-light company in France. The advantages of peat for the production of gas are as follows: 1st. It is less expensive than gas from coal, oil, or resin. 2d. The product is nearly as much as from those substances. 3d. The gas is quite harmless and inoffensive, and has, in respect to healthfulness, great advantages over some of the other kinds of gas. After it has been employed for gas the coke may be used for fuel, and is equal to any charcoal.

"If the above details may be relied upon, and if the experiments of Merle are satisfactory, if peat can be employed to advantage in the production of gas, it becomes one of the most important natural productions in the State, second only to coal for fuel, and equal to it for producing a beautiful and agreeable light. It would at once become a source of individual wealth, and furnish employment for a multitude of laborers, and increase the amount of transportation from the interior of the State to the cities and larger villages. It would employ a vast amount of material lying useless and unproductive, and one, too, embraced in our own territory. It would be using a great capital which has been accumulating for a long time, and has been reserved in store for this age of enterprise."

Dr. A. A. Hayes, of Boston, gives the following report of test, made by him, of peat sent to him from Wisconsin, as a material for the manufacture of illuminating gas: —

"The sample weighed eighteen pounds only, and was therefore used with other coal, substituting peat for a portion of the charge. One hundred and fifty pounds of coal make a charge for one retort.

"One hundred and thirty-four pounds of Pictou coal and sixteen pounds of Wisconsin peat were taken, and afforded —

Coke	101 lbs.
Gas	620 cub. ft.
Ammonia water and tar	———

"The gas had, after perfect purification, a spec. grav. 0.538, and measured 620 cubic feet. 100 parts contained olefiant gas 11.85. 5 cubic feet afforded as much light as 29 4.10 candles of standard spermaceti.

"134 lbs. Pictou coal would have afforded 536 cubic feet of gas; 5 feet being equal to 17 candles, or the whole to 1779 candles.

"134 lbs. Pictou coal added to 16 lbs. Wisconsin peat afforded 620 cubic feet; 5 feet being equal to 29 4.10 candles, or the whole to 3645 candles.

"There are only two or three cannel coals known which afford so much illuminating material, placing this peat in the first class of gas materials.

"It exceeds all common cannels, and, of course, is far above any bituminous coal, and can be worked with *poor* coal to make *good* gas."

At the Portland (Me.) Gas Works experiments were made, and the gas produced was used for lighting the city one or two evenings, with highly satisfactory results; the light furnished being, as we understand, full as good, if not better, than the quality ordinarily furnished from the coals which it is their custom to use.

The results of four experiments may be briefly stated as follows, it being understood that a given quantity of Albert coal was used in the retorts with the peat, as it is their custom to use it with the poorer coals from which most of their supply is obtained: —

1st.	720 lbs. Peat 100 " Albert	1st hour	1450 ft.	equal to	28	candles.
		2d "	1250 "	" "	19	"
		3d "	160 "	" "	15	"
2d.	1080 " Peat 270 " Albert	1st "	3190 "	" "	27	"
		2d "	2000 "	" "	18	"
		3d "	510 "	" "	10½	"
3d.	1080 " Peat 270 " Albert	1st "	2910 "	" "	24	"
		2d "	1960 "	" "	20	"
		3d "	480 "	" "	9	"
4th.	1080 " Peat 270 " Albert	1st "	2830 "	" "	25	"
		2d "	1810 "	" "	17	"
		3d "	350 "	" "	7	"

A gentleman connected with one of the large manufacturing establishments in Utica, N. Y., writes us as follows: "I have obtained from 160 lbs. of your condensed peat 612 cubic feet of illuminating gas. Our retorts were not well adapted to give the article a thorough test, and therefore am unable to give you a statement of its comparative cost."

Another correspondent writes, in Dec., 1865, "A piece of dried peat, containing less than six cubic inches, distilled in a small iron retort, with pipe-stem attached to the discharge pipe, furnished, for 33 minutes, a beautiful jet of pure clear gas, inodorous without purification, the illuminating power of which *appeared* (without actual test) to be greater than that we have from a single burner with ordinary gas."

Experiments were recently made at the Lansingburg (N. Y.) Gas Works. The peat used was simply dried in the sun, without pressing, and was then placed in the retorts. The gas was pronounced to be in every way superior to that made from the best coal. It gave a whiter, clearer, and much stronger light than the gas ordinarily produced there from coal, and stood the chemical tests well.

Professor Johnson, in his recent work, remarks, —

"It is essential that well-dried peat be employed. The retorts must be of good conducting material, therefore cast iron is better than clay. They are made of the ⌓ form, and must be relatively larger than those used for coal. A retort of two feet width, one foot depth, and 8 to 9 feet length, must receive but 100 lbs. of peat at a charge.

"The quantity of gas yielded in a given time is much greater than from bituminous coal. From retorts of the size just named 8000 to 9000 cubic feet of gas are delivered in 24 hours. The exit pipes must, therefore, be large, not less than 5 to 6 inches, and the coolers must be much more effective than is needful for coal gas, in order to separate from it the tarry matters.

"The number of retorts requisite to furnish a given volume of gas is much less than in the manufacture from coal. On the other hand, the dimensions of the furnace are considerably greater, because the consumption of fuel must be more rapid, in order to supply the heat which is carried off by the copious formation of gas.

"Gas may be made from peat at a comparatively

low temperature, but its illuminating power is then trifling. At a red heat alone can we procure a gas of good quality.

"The chief impurity of peat gas is carbonic acid: this amounts to 25 to 30 per cent. of the gas before purification, and if the peat be insufficiently dried, it is considerably more. The quantity of slaked lime that is consumed in purifying is therefore much greater than is needed for coal gas, and is an expensive item in the making of peat gas.

"While wood gas is practically free from sulphur compounds and ammonia, peat gas may contain them both, especially the latter, in quantity that depends upon the composition of the peat, which, as regards sulphur and nitrogen, is very variable.

"Peat gas is denser than coal gas, and therefore cannot be burned to advantage except from considerably wider orifices than answer for the latter, and under slight pressure.

"The above statements show the absurdity of judging of the value of peat as a source of gas by the results of trials made in gas works, arranged for bituminous coal.

"As to the yield of gas we have the following data, weights and measures being English:—

	Cubic ft.
100 lbs. of peat of medium quality from Munich, gave REISSIG .	303
" air-dry peat from Biermoos, Salzburg, gave RIEDINGER	305
" very light fibrous peat gave REISSIG	379 to 430
" Exter's machine-peat, from Haspelmoor, gave . . .	367

"Thenius states that, to produce 1000 English cubic feet of purified peat gas, in the works at Kempten, Bavaria, there are required in the retorts 292 lbs. of peat. To distil this, 138½ lbs. of peat are consumed in the fire; and to purify the gas from carbonic acid, 91½ lbs. of lime are used. In the retorts remain 117 lbs. of peat coal, and nearly 6 lbs. of tar are collected in the operation, besides smaller quantities of acetic acid and ammonia.

"According to Stammer, 4 cwt. of dry peat are required for 1000 cubic feet of purified gas.

"The quality of the gas is somewhat better than that made from bituminous coal."

Quite a number of experiments have been made with peat during the past year, in this country, for the purpose of testing its value for the production of illuminating gas; but we are not possessed of sufficient data concerning them to give any thing like accurate or reliable reports. The opinion is, however, freely expressed by those who have given most attention to it, that peat is destined at no distant day to be used very extensively for this purpose, both on account of the quality and quantity of the gas produced, and the low cost of the material, when compared with the coals at present most in use.

Peat in Gunpowder and Fireworks.

For the production of gunpowder, many varieties of peat are superior to the charcoal of dogwood and alder. We have seen the black peat of Massachusetts so perfectly prepared and granulated, without any explosive admixture, that it was impossible to distinguish it from the best rifle powder, even by a well-practised eye.

In the manufacture of fireworks, also, it is reported to have been used with marked success, from the fact that combustion is even more instantaneous and perfect than from the materials ordinarily in use, and that the fires produced exceed in brilliancy any thing heretofore attained.

Mr. J. B. Hyde, C. E., states that, "It has long been used by pyrotechnists in Europe in the composition of fireworks, particularly for colored fires, giving them greater brilliancy than could be effected by any other carbon.

"The artificer of Vauxhall Gardens, London, who declares it to be twenty per cent. more combustible than any other charcoal, long retained the secret of the material used by himself, which gave his works a wide reputation for their superiority."

Chemical Products from the Distillation of Peat.

In 1849, public attention in England was very much excited by an extraordinary statement made in the House of Commons by Mr. Mahon, and supported by Lord Ashley, asserting the immense area of bog in Ireland to contain substances of great economic value, to be produced at comparatively small cost. Lord Ashley's statements were based upon the authority of Mr. Owen, whose experiments went to show that, chemically treated at remunerative cost, the product would be the following substances: carbonate of ammonia, soda, vinegar, naphtha, paraffine, camphene oil, common oil, gas of value, and ashes.

The publication of this remarkable statement was quickly followed by a letter in the "Times," from Mr. Henry Seaman, of Plymouth, asserting that he and his neighbors lost £20,000 in an attempt to turn the peat-bogs of Dartmoor to profitable account, in the same manner as the peat of Ireland had been treated by Mr. Owen and his partner.

This letter was followed by one from Mr. Robert Oxland, a practical chemist, residing at Plymouth, who rather confirmed the principal parts of the statement made by Lord Ashley.

Subsequently the whole matter was taken up and thoroughly dissected in an article in the "Illustrated London News," No. 384, from which we quote a single paragraph: "No doubt a fair marketable value is taken for the several items; but we believe, and we express this most conscientiously, that the cost of production would exceed their commercial worth."

Again: in November, 1850, in the "Times," is found the following statement: "It now appears that Mr. Owen, whose course, from the first, was in no way inconsistent with Lord Ashley's testimony respecting him, has been, for the past year and a half, quietly engaged in testing the merits of the process to an extent that would properly authorize a definitive estimate of its results. These labors have been carried on partly under the superintendence of Dr. Hodges, the Professor of Agriculture in Queen's College, Belfast, and partly in the neighborhood of London, at the premises of Messrs. Coffey & Sons, engineers; and the conclusions now represented to have been arrived at, are of an exceedingly satisfactory nature. They do not promise the five hundred per cent. originally talked of, but according to a certified estimate rendered by Messrs Coffey & Sons, they show a profit of upwards of one hundred per cent.

"This estimate, which is framed for an establishment consuming 36,500 tons of peat per annum, is as follows: —

Expenditures.

"36,500 tons of peat, at 2s. per ton	£3650
455 " sulphuric acid, at £7	3185
Wear and tear of apparatus, &c.	700
Wages, labor, &c.	2000
Cost of sending to market, and incidentals . .	2182
Profit	11,908
	£23,625

Produce.

365 tons sulphate of ammonia, at £12 . .	£4380
255 " acetate of lime, at £14	3570
19,000 gallons naphtha, at 5s.	4750
109,500 lbs. paraffine, at 1s.	5475
73,000 gallons volatile oil, at 1s.	3650
36,000 " fixed oil, at 1s.	1800
	£23,625"

After referring again to the experience of Sir Robert Kane, as given in his "Industrial Resources of Ireland," and the experiments made by the Dartmoor Company on a large scale, the "Times" says, "We wish these results may be realized; but we have no hope of any thing so satisfactory."

Professor Brande read before the Royal Institution, in 1851, a paper upon peat and its products.

Special mention was made of the "tallow-peat" of the banks of Lough Neagh, which, from the brilliant flame attending its combustion, has been used for illuminating purposes as well as for fuel.

He says, "Peat may be rendered valuable either from the charcoal which may be obtained from it, or by various products derivable from what is called its destructive distillation.

"When it is desired to convert peat into charcoal, the plan adopted by the Irish Amelioration Society is to carbonize blocks of peat, partially dried on trays of wicker-work, in movable pyramidal furnaces. The charcoal so obtained varies in character with that of the peat which produces it; and, when the peat is compressed previous to its carbonization, the resulting charcoal exceeds the density of common wood charcoal.

"The efficacy of this charcoal in the manufacture of iron, in consequence of the small quantity of sulphur it contains, was mentioned, and its deodorizing and purifying qualities experimentally exhibited.

"The elements of peat are essentially those of wood and coal; viz., carbon, nitrogen, hydrogen, and oxygen."

In conclusion, Professor Brande reviewed the various products of peat derived by distillation, and their uses. They appear to be, —

1. *Sulphate of ammonia.* This substance is employed in the preparation of carbonate and muriate of ammonia, of caustic ammonia, and in the manufacture of fertilizing composts.

2. *Acetate of lime,* which is in constant demand as a source of acetic acid, and of various acetates largely consumed by calico printers.

3. *Pyroxylic spirit* (or wood alcohol), used in vapor-lamps, affording a brilliant light, and for the preparation of varnishes.

4. *Naphtha,* used for making varnishes, and for dissolving caoutchouc.

5. *Heavy and fixed oils,* applicable for lubricating machinery, especially when blended with other unctuous substance, or as a cheap lamp-oil, and as a source of lampblack.

6. *Paraffine.* This article is largely used for the manufacture of candles.

An article in the "Annual of Scientific Discovery," for 1851, relating to improvements in treating peat in England, says, "In addition to gas and ammonia, there is also obtained from the distillation of peat a peculiar acid, and a bitumino-adipose compound, which is called 'paranaphthadipose.' One of the products of this is a good solvent of gutta-percha, caoutchouc," &c.

The crude residues from the rectification of the oils of peat and bitumen, if burned in proper apparatus, furnish abundance of lampblack.

A writer in the "London Journal of Arts," for December, 1855, states that the peat-charcoal which is left in the retorts, after all the volatile constituents of peat have been distilled away, possesses the property of depriving colored vegetable solutions of the whole of their coloring matters. He adds that twenty-five per cent. more of this charcoal is needed than of bone-black; but the latter is about six times as expensive. Before using the peat-charcoal, it must be purified from iron, and sulphate of lime, and all alkaline matters.

In 1854 a company was formed, called the Irish Peat Company, having a factory near Athy, in Kildare, Ireland; its purpose being to produce, from peat, tar, paraffine, oil, naphtha, sulphate of ammonia, and to manufacture iron, the furnaces being heated with the gas manufactured or produced from the peat. It was under the general management and superintendence of Dr. Sullivan. The peat was distilled in furnaces, like the ordinary blast furnaces, thirty-two feet seven inches high, made perfectly tight by being incased in boiler plate iron, and covered at the top with a close conical valve and a double hopper. Air was blown in, in limited quantity, through three tuyères at the base. The volatile products were taken off at the top by two twelve-inch pipes, and conveyed into an hydraulic main three feet in diameter, from which the tar and other liquids flow into a tank, and the gases and vapors through series of condensing and purifying pipes and other apparatus, in which their separation is effectually completed. The charcoal is entirely consumed in the furnace. Various experiments gave great encouragement to the company, and the projectors, who were men of considerable standing and eminence, were sanguine of success. They started their works, however, at immense cost, and with the intention of producing *everything* from peat, in paying quantities, and consequently failed. It was well known, moreover, that the original outlay for building, machinery, &c., was enormous (something like $300,000), and the peat operated upon was of an inferior quality: the result of the enterprise, therefore, is no criterion for the commercial results of a prudently managed establishment.

Probably one of the reasons why several of the peat enterprises which have been started have met with so little real success has been that they wanted to do *too much.*

If people will undertake to obtain in one establishment *all* the products capable of being derived from peat, the probabilities are that they will fail of profit, and make a loss; but an establishment devoted exclusively to the preparation of peat as fuel, leaving others to char or distil it, will probably succeed commercially.

Mr. B. H. Paul read to the British Association, in 1863, a paper "On the Manufacture of Hydro-carbon Oils, Paraffine, &c., from Peat."

The author described the results which had been obtained at some works erected under his direction at Stornway, on the Island of Lewes.

The peat of that locality was described as a peculiarly

rich bituminous variety of mountain peat, yielding from five to ten gallons of refined oils and paraffine from the ton. The results obtained at these works were contrasted with those obtained at the works of the Irish Peat Company, where the product of oil was not more than two gallons per ton. This difference in the produce was ascribed, in a great degree, to the improper mode of working adopted at the Irish works.

One of the most important points dwelt upon was the necessity of regarding the hydro-carbon oils and the paraffine as the only products that would afford a profit in the distillation of peat. Dr. Paul concluded his paper by expressing his opinion that, though the working of peat was surrounded by many serious difficulties, there was every reason to believe that such peat as that occurring in the Highlands of Scotland could be advantageously worked; and that, if the manufacture of oils were undertaken with earnestness and perseverance, it would become the means of greatly improving the condition of those parts of the country, and a fertile source of profit.

In a recent conversation with a gentleman of this city, a native of Germany, he gave a very interesting account of a large peat establishment near his father's residence, "as large," he remarked, "as Chickering's Factory," where the products above mentioned are produced on an extensive scale, yielding, according to his statements, large profits to the proprietors of the works. He mentions also another production realized at these works, of which we have nowhere seen mention in connection with peat. We allude to *aniline colors*, which he states are there produced from peat, though in very small quantities; but, owing to their exceeding richness and strength, are sold "so many drops, so many dollars," which, being interpreted, means one dollar per drop. The finest candles used in the Catholic service in that country are manufactured from paraffine extracted from peat.

Analyses and Properties of Peat.

The ultimate elements of peat are essentially those of wood and coal; viz., carbon, hydrogen, oxygen, and nitrogen. If, therefore, peat be distilled in close vessels, the resulting products must be those of a similar operation on coal or wood.

In this way, as stated in the preceding chapter, it has been made to yield ammonia, acetic acid, pyroxylic spirit, tar, naphtha, oils, and paraffine, — all of great value in the useful arts.

Reports of analyses of peats are somewhat numerous, though in our own country but few have thus far appeared.

Peat always contains earthy matter to some extent, according to its position, thickness of stratum, &c., though sometimes in quantities so minute as to be accounted for only by the accumulations from the minute dust of the air. In other cases the situation is such that large amounts of earth are occasionally washed in from surrounding hills or blown in from more exposed locations. In this respect, therefore, certain deposits of peat may differ materially from wood. When the peat is consumed, these substances are left in the form of ash, and according to the nature of the ingredients of which it is composed, the ash is found to differ in color from white to gray and ochrey red.

Professor Johnson, who has devoted a good deal of personal attention to the matter in connection with agricultural interests, made report, in 1858, of results obtained by him from upwards of thirty specimens procured from various parts of the State of Connecticut. The following brief resume of his more elaborate report will serve to show something of the great variety of composition of this material : —

	Organic Matter.	Inorganic Matter.		Organic Matter.	Inorganic Matter.
1.	60	40	17.	70	30
2.	90	10	18.	35	65
3.	95	5	19.	23	77
4.	96	4	20.	32	68
5.	98	2	21.	60	40
6.	64	36	22.	26	74
7.	83	17	23.	25	75
8.	59	41	24.	28	72
9.	86	14	25.	92	8
10.	82	18	26.	97	3
11.	47	53	27.	84	16
12.	93	7	28.	90	10
13.	90	10	29.	68	32
14.	91	9	30.	88	12
15.	90	10	31.	73	27
16.	26	74	32.	76	24

Analyses of four samples of peat from Virginia, as conducted by him in November, 1866, are reported as follows : —

	I.	II.	III.	IV.
Water . . .	20.00	20.00	Dry.	Dry.
Volatile matter	50.05	52.59	62.56	65.74
Coke	24.97	25.44	31.21	31.81
Ash	4.98	1.97	6.23	2.45
	100.00	100.00	100.00	100.00

Four samples from the Great Dismal Swamp, analyzed in January, 1867, by Prof. B. Silliman, yielded as follows : —

	I.	II.	III.	IV.
Water . . .	78.89	20.22	Dry.	10.00
Volatile matter	13.84	52.31	65.53	58.08
Charcoal . . .	6.49	24.52	30.82	28.59
Ash	.78	2.95	3.65	3.33
	100.00	100.00	100.00	100.00

The following analyses by Regnault and Mulder show the contents of carbon, hydrogen, and oxygen: —

Locality.	Carbon.	Hydrogen.	Oxygen.	
Vulcaire . . .	59.57	5.96	34.47	Regnault.
	60.40	5.86	33.64	Mulder.
Long . . .	60.06	6.21	33.73	Regnault.
	60.89	6.21	32.90	Mulder.
Champ de Feu	60.21	6.45	33.34	Regnault.
	61.05	6.45	32.50	Mulder.
Freisland . .	59.42	5.87	34.71	"
Freisland . .	60.41	5.87	34.02	"
Holland . . .	59.27	5.41	35.35	"

The fact that a greater or less quantity of nitrogen is invariably contained in peat, appears to have been entirely overlooked in this case.

The "Dublin Journal of Industrial Progress" published an account of the results of analyses conducted by Drs. Kane and Sullivan, from which the following are selected, as indicating, apparently with great accuracy, the contents of carbon, hydrogen, oxygen, and nitrogen.

	Carbon.	Hydrogen.	Oxygen.	Nitrogen.
Surface Peat, Phillipstown . .	58.694	6.971	32.883	1.4514
Dense Peat " . .	60.476	6.097	32.546	.8806
Surface Peat, Bog of Allen .	59.920	6.614	32.207	1.2588
Dense Peat " " . .	61.022	5.771	32.400	.8070
Surface Peat, Twicknevin . .	60.102	6.723	31.288	1.8866
Surface Peat, Shannon	60.018	5.875	33.152	.9545
Dense Peat " . . .	61.247	5.616	31.446	1.6904

It should be observed that previous to analyses these samples were dried at a temperature of 220° Fahrenheit.

Further analyses, referring to the same subject, conducted by Kane, Ronalds, and Woskressensky, are given by Muspratt, as follows :—

	Carbon.	Hydrogen.	Oxygen.	Nitrogen.
Peat from Westmeath .	61.040	6.670	30.470	
" " Clare	56.630	6.330	34.480	
" " Kildare . . .	51.050	6.850	39.550	
" " Tuam	57.207	5.655	28.949	3.067
" " " . . .	58.306	5.821	29.669	2.509
" " " . . .	59.552	5.502	28.414	1.715
" " East'n Russia	39.084	3.788	51.088	

The following table of analyses of peat, both in the raw and carbonized state, from various localities, is found in Taylor's "Statistics of Coal :"—

By whom analyzed.	Carbon, per cent.	Vaporizable Matter, per cent.	Cinder, per cent.	Locality and Description.
David Mushet, Esq.	25.20	72.60	2.20	Scotch Peat, raw.
Dr. Kane	61.04	37.53	1.83	Bog of Allen, Ireland.
M. Marcher	65.00	22.00	13.00	Carbonized Peat.
	37.00	48.00	15.00	Raw Peat.
M. Debette	67.00	30.00	3.00	Bohemia, carbonized.
M. Berthier	38.40	28.00	17.40	Carbonized.
	24.40	70.60	5.00	Wurtemberg, raw.
Dr. C. T. Jackson .	21.00	72.00	7.00	Maine, U. S., raw.
M. Savage	22.00	69.70	8.30	Ardennes, France, raw.
M. Diday	9.00	58.00	33.00	Basse Alps, " "

From numerous analyses of dried peat, by Regnault, Mulder, Kane, Sullivan, Ronalds, and others, the average of results may be stated to be about as follows : —

	Max.	Min.	Average.
Carbon	61.247	51.05	60
Hydrogen . . .	6.971	5.41	6
Oxygen . . .	35.35	30.47	33
Nitrogen . . .	3.067	0.8070	1=100

Comparing these numbers with those afforded by dry wood, it would appear that an excess of ten per cent. of carbon and two per cent. of hydrogen is contained in the peat, over the quantity of these elements which wood ordinarily affords. This difference may have arisen from the decomposition which the matter of the peat has undergone.

"In general," says Sir Humphrey Davy, "one hundred parts of dry peat contain from sixty to ninety-nine parts of matter destructible by fire, and the residuum consists of earths, together with oxide of iron."

The following, which comprises a considerable variety, will indicate the percentage of ash remaining after peat is burned.

		Ash.	Observer.
Black firm peat from	Neumünster .	2.2	Suersen.
" " " "	Sindelfingen . .	7.2	Schubler.
Brown peat . . "	Schwenningen .	2.3	"
Old peat . . . "	Vulcare . . .	5.58	Regnault.
" " . . . "	Long	4.61	"
Peat . . . "	Champ de Feu	5.35	"
" . . . "	Berlin . . .	9.30	Achard.
" . . . "	" . . .	10.20	"
" . . . "	" . . .	11.20	"
Black old peat "	Maglin . . .	14.40	Eirchof.
Brown young peat "	—— . . .	14.30	"
Peat . . . "	Eichfield . .	21.50	Buchholz.
" . . . "	" . . .	23.—	"
" . . . "	" . . .	30.50	"
" . . . "	" . . .	33.—	"
Grass peat, light brown	. . .	1.5	Karmarsh.
Pitch peat	. . . . . .	8.—	"

Young peat, dark brown			. . .	7.—	Karmarsh.
Old peat			. . . .	10.—	"
41	kinds from	Erzgeberge	1.	@ 24.	Winkler.
3	" "	Holl.&Fries.	4.61	@ 5.580	Mulder.
27	" "	Bog of Allen	1.120	@ 7.898	Kane & Sullivan.
3	" "	Tuam	3.695	@ 4.819	Ronalds
9	" "	Saxony	5.300	@ 3.710	Wellner.

From this it will be seen that while some of these varieties are remarkably pure, in others no less than one third of the entire weight consists of incombustible substances, which, it is evident, is quite too large a quantity to admit of their being used as fuel. Such varieties are, however, valuable as manures, owing to the fact that they contain a large amount of phosphates and other salts, which serve to enrich the soil.

A knowledge of the composition of peat-ashes is of importance in determining their value for agricultural purposes, which is the use to which they are at present mainly applied, so far as we know, although it cannot be doubted that they possess a value for other purposes beyond what has thus far been developed.

Professors Kane and Sullivan, before referred to, are our authority for the analyses of nearly thirty specimens of ash, from which we take the following at random, as sufficient to illustrate the variety of their composition : —

	1	2	3	4	5	6	7	8	9	10
Specific Gravity	0.297	0.405	0.669	0.434	0.984	1.058	0.481	0.280	0.335	0.924
Potassa	0.362	1.323	0.461	0.641	0.347	0.744	1.667	0.146	0.491	0.280
Soda	1.427	1.902	1.399	1.875	0.679	0.704	2.823	0.466	1.670	2.180
Lime	26.113	36.496	40.920	22.702	45.581	40.623	20.907	8.492	33.037	30.744
Magnesia	3.392	7.634	1.611	6.809	1.256	4.352	15.252	4.702	7.523	9.237
Alumina	4.180	5.411	3.793	1.109	0.129	1.671	2.034	10.705	1.686	2.027
Sesquioxide of Iron	11.591	15.608	15.969	29.845	15.974	10.368	17.040	15.052	13.281	19.797
Phosphoric Acid	1.461	2.571	1.406	2.019	0.188	1.114	1.447	1.557	1.438	1.290
Sulphuric Acid	12.403	14.092	14.507	16.381	44.371	24.208	23.375	13.974	20.076	20.857
Hydrochloric Acid	1.568	1.482	0.983	1.591	0.337	1.052	1.424	0.196	1.747	3.128
Silica in Compounds decomposable by Acids	0.980	3.595	1.111	0.737	1.043	6.317	6.634	12.476	2.148	3.096
Sand and Silicates undecomposable by Acids	22.519	2.168	2.107	14.505	2.653	3.710	10.682	31.198	7.683	3.163
Carbonic Acid	13.695	7.761	15.040	1.470	16.120	4.981	6.721	...	8.340	3.570
Total	99.691	100.043	99.307	99.693	99.678	99.844	100.006	98.928	99.120	99.369

As to the chemical products of peat, a series of experiments was instituted in the laboratories of the Museum of Irish Industry, under the superintendence of Sir Robert Kane and Professor Sullivan.

These were conducted upon specimens of peat, from various peat deposits in Ireland, in a two-fold manner.

First, by distilling the peat in close vessels, and secondly, by effecting the distillation by a combustion of a part of the material with a blast of air.

In operating by the former method, a retort was used similar to that employed in the distillation of coal in the manufacture of gas. To the exit pipe from this retort, a series of Wouffe's bottles were adapted, wherein the greater portion of the tar and other aqueous products were deposited, the remainder being condensed in a worm fixed in a barrel of water, and the permanent gases collected. In each trial one hundred pounds of wet peat was worked off in eight to fourteen charges, according to the density of the substance.

The following table represents the quantity of the gross products obtained as the result of seven experiments: —

	Water.	Tar.	Charcoal.	Gas.
1.	23.600	2.000	37.500	36.900
2.	32.273	3.577	39.132	25.018
3.	38.102	2.767	32.642	26.489
4.	38.628	2.916	31.110	32.346
5.	32.098	2.344	23.437	42.121
6.	38.127	4.417	21.873	35.693
7.	21.189	1.462	18.973	57.746
Average . .	31.378	2.787	29.222	36.616

The amount of ammonia, acetic acid, and naphtha contained in the aqueous product was next determined, as well as that of the oil and paraffine in the tar, and the following table contains the results obtained : —

	Ammonia.	Acetic Acid.	Pyroxylic Spirit, or Naphtha	Paraffine.	Volatile Oil.	Fixed Oil.
1.	0.302	0.076	0.092	0.024	0.684	0.469
2.	0.187	0.206	0.171	0.179	0.721	0.760
3.	0.393	0.286	0.197	0.075	0.571	0.565
4.	0.210	0.196	0.147	0.170	1.262	0.617
5.	0.195	0.208	0.161	0.196	0.816	0.493
6.	0.404	0.205	0.132	0.181	0.829	0.680
7.	0.181	0.161	0.119	0.112	0.647	0.266
Average. .	0.268	0.191	0.146	0.134	0.790	0.550

In the second series of experiments in which the peat was consumed by igniting it, and then supporting the combustion by a blast, the following results were obtained from the same peats used in the former experiments, selecting from them those kinds which had shown the greatest dissimilarity.

	Water.	Tar.	Ash.	Gases.
1. . . .	31.678	2.510	2.493	63.319
2. . . .	30.663	2.395	7.226	59.716
3. . . .	29.818	2.270	2.871	65.041
Average .	30.720	2.392	4.197	62.692

The amount of ammonia, acetic acid, naphtha, paraffine, and oils, obtained in the same way, were as follows : —

	Ammonia.	Acetic Acid.	Naphtha.	Paraffine.	Oils.
1. . .	0.322	0.179	0.158	0.169	1.220
2. . .	0.344	0.268	0.156	0.086	0.946
3. . .	0.194	0.174	0.106	0.119	1.012
Average	0.287	0.207	0.140	0.125	1.059

The average results of both methods of operation gave for the distillation in close vessels, —

	Maximum.	Minimum.	Average.
Aqueous products . . .	31.678	29.818	30.614
Tar	2.510	2.270	2.392
Ashes	7.226	2.493	4.197
Gases	65.041	59.716	62.392

The watery products and tar afforded, were, —

	Maximum.	Minimum.	Average.
Ammonia	0.344	0.194	0.287
or as Sulph. of Ammonia .	1.330	0.745	1.110
Acetic Acid . . .	0.268	0.174	0.207
or as Acetate of Lime . .	0.393	0.256	0.305
Naphtha	0.158	0.106	0.140
Paraffine	0.169	0.086	0.125
Volatile and Fixed Oils	1.220	0.946	1.059

The relative value of the two methods is as follows : —

	Average product from close distillation.	Average product from distillation in blast of air.
Ammonia	0.268	0.287
or as Sulph. of Ammonia	1.037	1.110
Acetic Acid	0.191	0.207
or as Acetate of Lime	0.280	0.305
Naphtha	0.146	0.140
Paraffine	0.134	0.125
Oils	1.340	1.340

The similarity of these numbers shows that very little difference exists between the results obtained from the distillation of the substance in close vessels, and from its decomposition in the fire by a blast of air. In the first case, however, there remains an average of twenty-nine parts of charcoal, which might determine a preference for that method were it not that the greater part of this is consumed in carbonizing the matter in the retort.

A similar and independent series of experiments was conducted by Dr. Hodges, whose results, as contrasted with those arrived at by Professors Kane and Sullivan, we are enabled to give in connection with the statement of products anticipated by the Irish Peat Company, before alluded to, and which formed the basis of their operations and calculations of gain.

	Kane and Sullivan.		Hodges.		Irish Peat Co. Prospectus.	
	Per ct.	Per ton.	Per ct.	Per ton.	Per ct.	Per ton.
Sulph. Ammo.	1.110	$24\frac{3}{10}$ lbs.	1.000	$22\frac{3}{4}$ lb.	1.000	$22\frac{2}{3}$ lb.
Acetic Acid	0.207	$4\frac{2}{3}$ lbs.	0.328	$7\frac{1}{4}$ lb.	..	..
or as						
Acetate of Lime	0.305	$6\frac{4}{5}$ lbs.	..	..	0.700	$15\frac{7}{10}$ lb.
Naphtha	0.140	$50\frac{1}{5}$ oz.	0.232	$83\frac{1}{4}$ oz.	0.185	$66\frac{3}{10}$ oz.
Tar	2.392	$53\frac{3}{4}$ lbs.	4.440	$99\frac{1}{2}$ lb.	..	..
Products of the Tar. Paraffine	0.125	$2\frac{4}{5}$ lbs.	..	..	0.104	3 lb.
Products of the Tar. Oils . . .	1.059	$2\frac{2}{3}$ gals.	..	..	0.071	$2\frac{1}{2}$ gal.

These results show a good degreee of similarity, especially when it is considered that there was a very considerable variety in the quality and characteristics of the samples experimented upon.

The heating power of peat turf, dried but not condensed, and peat charcoal, as estimated with reference to the heating power of carbon, which is taken as unity, is thus stated in "Scheerer's Metallurgie": —

Imperfectly air-dried turf, with 30 per cent. of hygroscopic moisture and 10 per cent. of ash,	0.37
Best air-dried turf, with 25 per cent. of moisture and no ash, ...	0.47
Kiln-dried turf, with no moisture and 15 per cent. of ash, ...	0.55

Best kiln-dried turf, without moisture or ash, . .	0.65
Poorest kind of air-dried peat charcoal, with 10 per cent. moisture and 56 per cent. ash, . .	0.85
Best air-dried peat charcoal, with 10 per cent. moisture and 4 per cent ash, ...	0.83

With a view to ascertain its relative calorific power, Mr. C. Cowper, of London, experimented by the "litharge test," recommended by Berthier, upon peat from the Bog of Allen. This test consists in mixing a given weight of fuel with a sufficient quantity of litharge, and heating it in a crucible: the heating power is in proportion to the quantity of lead reduced. By his experiments, the following comparative results were obtained, being averages of six or eight experiments each:—

10 grains	of good Newcastle coal gave . .	284	grains.
10 "	" oven coke gave	302	"
10 "	" common peat (unmanufactured) gave	144	"
10 "	" same, coked in a crucible, gave .	259	"

The foregoing comparison is founded upon a well-known fact, that the quantity of heat generated during the combustion of any fuel is in exact relation to the quantity of oxygen consumed in the process. Hence, in order to ascertain the relative calorific power of different kinds of fuel, it is only necessary to ascertain the quantity of oxygen which each consumes in burning.

These experiments show that seven tons of coke from crude unmanufactured peat were equal to six tons of good coal coke; but the real value of the experiments and their results are very much lessened from the fact stated, and for which no reason is given, that the peat used was "of an inferior quality."

Professor Everitt conducted similar experiments, with results as follows: —

10 grains peat coke, picked surface, gave	. 277	grains.
10 " " " lower strata, gave .	. 250	"
10 " pressed peat gave	. 137	"

Investigations conducted by Berthier, Griffiths, and Winkler, upon the same principle, are reported as follows: —

	Pounds of lead reduced by one lb. peat.	Pounds of water heated from 32° to 212° by one lb. peat.	
Peat from Troyer	8.0	18.1	Berthier.
" " Ham	12.3	27.9	"
" " Passy	13.0	29.2	"
" " Framont	15.4	34.9	"
" " Ischoux	15.3	34.6	"
" " Konigsbrunn . . .	14.3	32.4	"
" " Bog of Allen . . .	27.7	62.7	Griffiths.
" " " " "	25.0	56.6	"
" " among 24 kinds from Hartz Mountain, the worst gave .	11.9	26.9	Winkler.
" " the best gave . . .	18.8	42.6	"
Peat charcoal from Seine . . .	17.7	40.1	Berthier.
" " " Ham . . .	18.4	41.7	"
" " " Essone . .	22.4	50.7	"
" " " Framont . .	26.0	58.9	"

Peats differ as much in specific weight as in composition. The cause of the difference is, perhaps, not so clearly ascertained, but it is supposed that the degree of decomposition which the substance has undergone determines principally, in most cases, the difference of specific weight in peat from the same locality.

Professor Everitt's investigations of the common

Lancashire peat show that, in regard to comparative specific gravity, —

Compressed peat possesses	1.160
Less pressed peat possesses	.910
Peat coke, hard pressed, possesses	1.040
Peat coke, less pressed, possesses	.913
Charcoal from hard woods, possesses	.400 to .625

Hence it appears that the coke prepared from compressed peat is nearly double the density of ordinary charcoal. In common practice, it has been the custom to estimate that one hundred pounds of charcoal occupy the same space as two hundred pounds of coke. Peat coke would occupy, weight for weight, the same space as common coke.

Professor Everitt adds that, where bulk of stowage and high intensity of heat are important considerations, the peat coke is superior to charcoal. Moreover, the density of peat coke may, by means of improvements in the mode of manufacture recently introduced, be carried up from thirty to fifty per cent. beyond that indicated above; which, of course, increases the comparison in its favor.

The calorific value or power of peat coke may be commercially averaged as equal to coal coke. The evaporating powers of the two are nearly equal; but peat coke has the advantage of *freedom from sulphur*, and those conversant with the use of fuel in the manufacture of metals will readily appreciate this.

Its superiority is decided when used for the following purposes: —

For the working of malleable iron.

For melting unmalleable or cast iron.

For the smelting and general manufacture of iron from the ore; and

For all descriptions of brass and copper work.

Karmarsh arrived at the following results in regard to the specific weights of Hanoverian peat, to wit: —

Light colored young peat, nearly unchanged moss	0.113 to 0.263
Young brownish black peat, an earthy matrix intersected with roots	0.240 to 0.600
Old earthy peat, without any fibrous texture.	0.564 to 0.902
Old or pitch peat	0.639 to 1.039

Of a large number of samples examined by Sir Robert Kane and Dr. W. K. Sullivan, the results, as shown below, serve to indicate the wide difference which is found to exist in this respect in European peats: —

1. . 0.297	5. . 0.351	9. . 0.655	13. . 0.523	17. . 0.323	21. . 0.629
2. . 0.405	6. . 0.661	10. . 0.434	14. . 0.274	18. . 0.924	22. . 0.280
3. . 0.669	7. . 0.335	11. . 0.984	15. . 0.394	19. . 1.058	23. . 0.546
4. . 0.450	8. . 0.476	12. . 0.681	16. . 0.437	20. . 0.481	24. . 0.855

The maximum showing a density of 1.058, and the minimum 0.235.

Similar results may be said to characterize the peats of this country, varying, as there, according to the age, location, climate, and character of the deposits.

Tests, Experiments, and Testimony.

We had intended to devote a chapter especially to statements relating to trials of this fuel which have been made, as demonstrating the practicability of using it, its relative value, and the great variety of purposes for which it may be easily, effectually, and economically used; also statements and opinions of practical and scientific men, which would be of interest, and are entitled to consideration in this connection.

So much testimony and so many statements of this character have, however, been incorporated in what we have said in preceding chapters, that more are not required, and it is deemed sufficient simply to indicate on what pages those already recited may be found, to wit:

Miscellaneous, — Pages 32, 59, 67, 71, 72, 74, 82, 83, 84, 85, 86, 87, 89, 98, 99, 100, 103, 110, 112, 114, 115, 116, 120, 121, 122, 123, 141, 147, 148, 149, 162, 165, 166, 170.

Peat Charcoal. — 70, 71, 72, 243, 244, 250, 254, 256, 257, 258, 259, 260.

Working Iron. — 174, 175, 176, 177, 178, 179, 180, 181, 182, 183, 184, 185, 188, 189, 190, 191, 192, 193, 194, 195, 196, 197.

Generating Steam. — 198, 199, 200, 201, 202, 203, 204, 205, 206, 208, 209, 210, 211, 212, 213, 214, 215, 216, 225, 226, 227, 228.

Domestic Purposes. — 217, 219, 220.

Intensity of Heating Power. — 71, 72, 222, 257, 258, 259, 260.

How to use Peat. — 225, 226, 227, 228.

Gas. — 166, 230, 231, 232, 233, 234, 235, 236, 237, 238, 239.

Gunpowder and Fireworks. — 240.

Chemical Products. — 240, 241, 242, 243, 244, 245, 246, 254, 255, 256, 257.

Analyses and Properties. — 248, 249, 250, 251, 252, 253, 254, 255, 256, 257.

Specific Gravity. — 259, 260, 261.

As before remarked, it should be borne in mind that peats differ in quality and characteristics, and consequently in their heating properties and value, full as much as wood and coal in all their varieties; moreover, that the results obtained from this fuel, as from any other, will in all cases depend very much upon the manner in which it is burned. We are told that a certain service is obtained from a boiler with one cord of wood, or one ton of coal; but unless we know whether the wood be pine or hickory, or the coal Pictou or Lehigh, we have no certain data from which to make accurate or even approximate calculations. So also of peat; it is necessary to know somewhat of its characteristics in each case reported, in order to form a correct estimate of its relative value for the service done. The terms by which to designate these characteristics and qualities, so as to be generally understood and adopted, have yet to be determined upon; but the necessities of the case will doubtless bring them out in good time.

Peat for Pavements.

The solid bitumen from the distillation of peat may be employed like asphalt in the preparation of mastic for paving; and experiments have shown that peat itself may be converted into a similar material by the following process: Having been well dried, it is mingled with from ten to fifteen per cent. of coal-tar, and the mixture boiled for several hours, until the peat dissolves into a viscid liquid, which, when cooled, is solid, and resembles asphalt.

Another account says, —

"For *pavements*, when combined with an artificial asphaltum composed of carbonate of lime and coal-tar, it is said to form a solid and elastic road, superior, in many respects, to native asphaltum. The tendency of this artificial asphalt to crack and break is counteracted by the strong fibre of the peat; which, if added to the chalk and tar while warm, acts as a binder when the mass is cooled, and obviates its brittleness."

Paper from Peat.

J. Lallemand, of Besançon, France, procured, in 1854, a patent for producing paper from the fibrous portions of peat, mixed with from five to ten per cent. of rag pulp, the peat having been treated by a peculiar process described by him; after which, the mixture is subjected to the ordinary processes for making paper.

Experiments on a limited scale, in this country, have demonstrated the practicability of producing good paper from some kinds of peat.

Peat for Building and Ornamental Work.

During the past season we have several times suggested that the characteristics of peat were such that it required but little ingenuity to make use of it for building, ornamental, and other purposes, in the same manner as, or as a substitute for, terra-cotta and papier-maché.

More recently a paragraph has been shown us, cut from the "Builder" (1860), stating that improvements in manufactures of peat had been patented by Mr. H. Hodgson, of Ballyreine and Merlin Park, and Mr. P. M. Crane, of the Irish Peat Works, Athy, which consist in preparing from peat, in its natural state, blocks, slabs, or pieces of any size, form, or thickness, which, when so prepared, are said to be useful and economical in the construction of parts of buildings, and for various other purposes.

These blocks are placed between cloths, of woven or textile fabric, or other suitable material; and the peat is placed between shelves, and submitted to hydraulic or other pressure. Part of the water is forced out, and the peat solidified; and drying is effected either by exposure to the atmosphere, or in a room heated artificially, or by any other process. They are then put again between the plates of an hydraulic or other press, and extreme pressure put on them.

If the product of this invention be required for use for inside work in building, such as partitions, linings, inside roofing, or for other work, as a non-conducting substance, they do not require other further preparation than shaping, provided they are not to be exposed to moisture. But the slabs or pieces used for roofing (instead of slates, tiles, or other things of that nature) are prepared to resist the wet or action of the atmosphere by steeping them in, or saturating or coating them with, some fitting material to resist moisture.

Toys, fancy articles, and even rings and jewelry, have been produced from peat, and have much the same appearance as the same articles made from India rubber.

Peat for Tanning Leather.

It has recently been stated, on what appears to be good authority, that the amount of resinous and vegetable matter in some peats render them apparently a superior material for tanning purposes.

If this should prove to be the case, it will require only to demonstrate and establish the fact, and a market will readily be created. Once adopted, the consumption of it would be large and steady. The matter is certainly worthy of careful investigation.

Antiseptic Properties of Peat.

We have had occasion several times to refer to the antiseptic properties of peat. On this subject, Professor Lyell, whom we have before quoted, says, —

"One interesting circumstance attending the history of peat-mosses, is the high state of preservation of animal substances buried in them for periods of many years. In June, 1747, the body of a woman was found six feet deep in peat, in a peat-moor in the Isle of Axholm, in Lincolnshire. Upon the feet were leathern shoes, or sandals, each cut out of a single piece of tanned ox-hide, folded about the foot and heel, and piked with iron. Such are described by Chaucer as being worn in his time. This certainly afforded evidence of her having been buried there for many ages; yet her nails, hair, and skin are described as having shown hardly any marks of decay. In a turbary on the estate of the Earl of Moira, in Ireland, a human body was dug up, a foot deep in gravel, covered with eleven feet of peat: the body was completely clothed, and the garments seemed all to be made of hair. Before the use of wool was known in that country, the clothing of the inhabitants was made of hair; so that it would appear that this body had been buried at that early period: yet it was fresh and unimpaired. In the Philosophical Transactions, we find an example recorded of the bodies of two persons having been buried in moist peat in Derbyshire, in 1674, about a yard deep, which were examined twenty-eight years and nine months afterwards. 'The color of their skin was fair and natural; their flesh soft as that of persons newly dead.'"

In Dr. Rennie's Essays, reference is made to several other instances of like character.

"Among other analogous facts, we may mention that, in digging a pit for a well near Dulverton, in Somersetshire, many pigs were found in various postures, still entire. Their shape was well preserved; the skin, which retained the hair, having assumed a dry, membraneous appearance. Their whole substance was converted into a white, friable, laminated, inodorous, and tasteless substance, but which, when exposed to heat, emitted an odor precisely similar to broiled bacon.

"We naturally ask whence peat derives this antiseptic property. It has been attributed by some to the carbonic and gallic acids which issue from decayed wood, as also to the presence of charred wood in the lowest strata of many peat-mosses; for charcoal is a powerful antiseptic, and capable of purifying water already putrid. Vegetable gums and resins also may operate in the same way. Dr. Macculloch suggests that the soft parts of animal bodies, preserved in peat-bogs, may have been converted into adipocere by the action of water merely.

"The Solway Marsh is a flat area, about seven miles in circumference, situated on the confines of England and Scotland. It is related that at the battle of Solway, in the time of Henry VIII. (1542), when the Scotch army, commanded by Oliver Sinclair, was routed, an unfortunate troop of horse, driven by their fears, plunged into this morass, which instantly closed upon them. The tale was traditional; but the fact that a man and horse in complete armor were actually found by peat-diggers in the place where it was always supposed the affair had occurred, tends to authenticate the story. The skeleton of each was well preserved, and the different parts of the armor were easily distinguished.

"This same moss, on the 16th of December, 1772, having been filled with water during heavy rains, rose to an unusual height, and then burst. A stream of black, half-consolidated mud began at first to creep over the plain, resembling, in the rate of its progress, an ordinary lava current. No lives were lost; but the deluge totally overwhelmed some cottages, and covered four hundred acres. The highest parts of the original moss subsided to the depth of about twenty-five feet, and the height of the moss on the lowest parts of the country which it invaded was at least fifteen feet." Several other instances of a similar character might be cited.

"The antlers of large and full-grown stags are among the most common and conspicuous remains of animals in peat. They are not horns which have been shed; for portions of the skull are found attached, proving that the whole animal perished. Bones of the ox, hog, horse, sheep, and other herbivorous animals, also occur; and in Ireland, and the Isle of Man, skeletons of a gigantic elk; but no remains have been met with belonging to those extinct quadrupeds of which the living congeners inhabit warmer latitudes, such as the elephant, rhinoceros, hippopotamus, hyena, and tiger, though these are so common in superficial deposits of silt, mud, sand, or stalactite, in various localities throughout Great Britain. Their absence seems to imply that they had ceased to live before the atmosphere of this part of the world acquired that cold and humid character which favors the growth of peat.

"We are informed by Deguer, that remains of ships, nautical instruments, and oars, have been found in many of the Dutch mosses; and Gerard, in his 'History of the Valley of the Somme,' mentions that in the lowest tier of that moss was found a boat loaded with bricks, proving that these mosses were, at one period, navigable lakes, and arms of the sea, as were also many of the mosses on the coast of Picardy, Zealand, and Friesland. The canoes, stone hatchets, and stone arrow-heads, found in peat, in different parts of Great Britain, lead to similar conclusions."

Professor Lyell, in a work on geology, published in London in 1865, says of the peat of Denmark, "In the lower beds of peat (a deposit varying from twenty to thirty feet in thickness), weapons of stone accompany trunks of the Scotch fir; while, in the higher portions of the same bogs, bronze implements are associated with trunks and acorns of the common oak. All the quadrupeds found in the peat agree specifically with those now inhabiting the same districts, or which are known to have been indigenous in Denmark within the memory of man."

Peat as a Disinfectant and Deodorizing Agent.

Peat is very commonly used as a disinfectant and deodorizer, and every farmer knows its efficacy in these respects, when placed about vaults and drains, or other receptacles of filth.

A famous "*Chemical Deodorizing Powder*," which has been extensively sold throughout the country for fifteen years past, is simply *peat*, charred, pulverized, and put up in neat packages convenient for use, which are inscribed as follows: —

"*Nature is ever true to Herself.* — This preparation is the greatest absorbent of carbonic acid gas in nature, and also of all those noxious and poisonous *miasma* which are generated in thickly populated districts, from the decomposition of animal and vegetable substances. It is a great antiseptic, and will prevent the cholera and fever from entering your premises, as it absorbs all those noxious malaria which are so prevalent during the warm season. It is indispensable in the sick room and in sleeping rooms, as a small quantity in an open vessel will keep the air pure and agreeable. It supersedes every other neutralizing substance, as it is entirely harmless and without odor of any kind. It is highly recommended by eminent chemists and medical men. *It is really the abater of every nuisance.*"

Although "nothing but peat," we are satisfied from our own experience, and testimony abundant, that the statement is true, and that it does possess the properties and virtues claimed for it.

PEAT AS A FERTILIZER.

Peat, as a fertilizer, possesses a value to our farming and agricultural interests beyond what is generally accorded to it. The "Muck Manual," by S. L. Dana, of Lowell, published in 1843, is considered standard authority upon this latter point, and will be consulted with interest and profit by our farmers, as will also "Johnson's Essays on Peat, Muck, and Commercial Manures," recently republished, with additions, under the title of "*Peat and its Uses as Fertilizer and Fuel*," * in which the author treats of the characteristics that adapt it for agricultural uses, as, for instance, its remarkable power of absorbing and retaining water, both as a liquid and as a vapor; its power of absorbing ammonia; its effect in promoting the disintegration and solution of mineral ingredients of the soil, and its influence on the temperature of the soil. He also treats of the ingredients and qualities which makepeat a direct fertilizer; which are, the organic matters, including nitrogen; the inorganic or min-

* Published by Orange Judd & Co., 41 Park Row, N. Y., 1866. $1.25.

eral ingredients ; also some peculiarities relating to the decay of peat. The whole subject is treated with care and in detail, and much valuable data is given for estimating and illustrating its practical value as compared with stable manure, guano, &c. Farmers ought to have the book, and we cannot do them better service than to advise them to *buy* it and *read* it.

Professor Dana says, "Peat is too well known to render it necessary to say that it is the result of that spontaneous change in vegetable matter which ends in geine. Peat is, among manures consisting chiefly of geine, what bone-dust is among manures consisting of animal matter. Peat is highly concentrated vegetable food. When the state in which this food exists is examined, it is found not only partly cooked but seasoned.

"Peat consists of soluble and insoluble geine and salts. The proportion of these several ingredients must be known before the value of peat can be compared with similar constituents in cow-dung. This proportion is exhibited in the following table of constitution of Massachusetts peats per 100 parts : —

Locality.	Soluble Geine.	Insoluble Geine.	Total Geine.	Salts and Silicates.
1. Dracut	14.	72.	86.	14.
2. Sunderland	26.	56.60	85.60	14.40
3. Westborough.	48.80	43.60	92.40	7.60
4. Hadley	34.	60.	94.	6.
5. Northampton	38.30	44.15	82.45	17.55
6. "	32.	54.90	86.90	13.10
7. "	12.	60.85	72.85	27.15
8. "	10.	49.45	59.45	40.55
9. "	33.	59.	92.	8.
10. "	46.	46.80	92.80	7.20
Average	29.41	55.03	84.44	15.55
11. Watertown, pond mud .	5.10	8.90	14.	86.
12. Danvers, pond mud . .	8.10	6.50	14.60	84.40

"Under the general name of *peat*, are comprised several varieties, which may be distinguished as, 1st. Peat, the compact substance generally known and used for fuel, under this name. 2d. Turf, or swamp muck, by which is to be understood the paring which is removed before peat is dug. It is a less compact variety of peat. It is common in all meadows and swamps, and includes the hassocks. Both these varieties are included in the foregoing, from No. 1 to No. 10. It includes also the mud of salt marshes. 3d. Pond-mud, the slushy material, found at the bottom of ponds when dry, or in low grounds, the wash of higher lands. This seldom contains more than 20 per cent. of geine. Nos. 11 and 12 are of this description.

"These varieties comprise probably a fair sample of all the peat and swamp muck and pond mud which occur in the various parts of the country. The results stated are those of the several varieties, when dried, at a temperature of 240° Fahr.

"Peat ashes contain all the inorganic principles of plants which are insoluble, with occasional traces of the soluble alkaline sulphates and of free alkali.

"All peat shrinks by drying, and when allowed to drain as dry as it will, it still contains about $\frac{2}{3}$ of its weight of water. It shrinks from $\frac{2}{3}$ to $\frac{3}{4}$ of its bulk. A cord wet becomes $\frac{1}{4}$ to $\frac{1}{3}$ of a cord when dry. To compare its value with cow-dung, equal bulks must be taken.

"It is found, on analysis, that this does not differ much from fresh cow-dung, so far as salts, geine, and water are concerned. The salts of lime are actually about the same, while the alumina, oxide of iron, magnesia, in the silicates, added to the salts of lime, make the total amount of salts, in round numbers, equal that of cow-dung."

He gives the following comparative weights and composition of a cord of cow-dung and a cord of each of two kinds of peat referred to in the foregoing table:—

	Weight.	Soluble Geine.	Insoluble Geine.	Total Geine.	Salts of Lime.
Dung	9289 lbs.	128 lbs.	1288 lbs.	1416 lbs.	92 lbs.
Peat, No. 9 . .	9216 "	376 "	673 "	1049 "	91 "
Peat, No. 10 . .	9216 "	519 "	529 "	1048 "	81 "

A cord of pond mud (No. 11) weighs, when dug, 6117 lbs., and contains solid matter 3495 lbs., composed of geine 495 lbs., of silicates and salts 3005 lbs. The salts of lime in pond mud are 2½ per cent.

The salts and geine of a cord of peat are equal to the manure of one cow for three months.

It is certainly a very curious coincidence of results, that Nature herself should have prepared a substance whose agricultural value approaches so near to cow-dung, the type of manures.

The power of producing alkaline action on the insoluble geine is alone wanted to make peat good cow-dung, and the question arises, How is to be given to peat (a substance which, in all its other characters, is so nearly allied to cow-dung) that lacking element, ammonia? How is that to be supplied? Without it, cow-dung itself would be no better than peat, not so good even; for in peat nearly one half of the geine is already in a soluble state. By the addition of *alkali* to peat, it is put into the state which ammonia gives to dung, and it is found that for all agricultural purposes the desired result is obtained by adding to every cord of fresh dug peat 90 to 100 lbs. pot or pearl ashes, or 60 to 65 lbs. of soda, or 16 to 20 bushels of common wood ashes.

Abundant testimony is afforded that a cord of clear stable manure composted with two cords of peat, forms a manure of equal value to three cords of green dung.

Ashes of Peat.

The ashes of peat are of very considerable value, are used to great advantage on some soils, and highly esteemed. Details of the composition of a variety of samples are given on page 253.

Professor Dana says, "Peat ashes abound in carbonate, sulphate, and especially phosphate of lime. Free alkali may always be traced in peat ashes; but alkali exists in it, rather as silicate, as in leached ashes."

A correspondent writes, "Ship-loads of peat ashes from Holland, for the use of the London market gardeners, are imported every year. They are grand for young clover."

They also make a very serviceable cement, and are used to some extent for that purpose. It has been suggested that they would make an excellent polishing powder for lithographic stones, metals, &c. They certainly possess a value, and should not be allowed to go to waste.

Conclusions.

In view of these statements, thus hastily given, and which have been extended beyond our original plan, although barely sufficient to introduce the subject, it must, we think, be evident to the most hasty reader, that the substance of which we treat is of sufficient importance to command earnest attention, not only from the business man, on the score of its application to domestic purposes, manufactures, and the arts, but from the philanthropist, in view of the relief it may be made to afford as one of the necessaries of life.

The facts narrated, aside from those derived from

personal experience and observation, are collected from a variety of sources; and, in the hope that others may be led to investigate and pursue the subject more at length, we add a list of authorities, from nearly all of which we have drawn something (using freely, in many cases, the exact language of the several writers in connection with our own), in many of which the subject is treated in detail, and in some with great ability.

Geological Surveys of Great Britain.
Martin's Statistics of the British Colonies.
Lindley and Hutton's Fossil Flora of Great Britain.
Macculloch's Statistics of the British Empire.
Macculloch's Geographical and Statistical Dictionary.
Sullivan's Journal of Industrial Progress.
Kane's Industrial Resources of Ireland.
Ireland and its Economy.
Wakefield's Account of Ireland.
Ordinance Survey, and report of the County of Londonderry.
Lyell's Principles of Geology.
Dana's Manual of Geology.
Jameson's Mineralogy of the Scottish Isles.
Dr. R. Rennie on Peat.
Ure's Dictionary of Arts, Manufactures, and Mines.
Appleton's Dictionary of Mechanics.
Muspratt's Chemistry as applied to Arts and Manufactures.
Encyclopædia Americana.
Encyclopædia Britannica.
New American Cyclopædia.
Brewster's Edinburgh Encyclopædia.
Rees' Cyclopædia.
Encyclopædia Pertheasis.
Encyclopædia Metropolitana.
Tomlinson's Encyclopædia.
Homan's Cyclopædia of Commerce.
Loudon's Encyclopædia of Agriculture.
Buckland on Coal.

Johnson on Coal.
Williams on Combustion of Coal.
Taylor's Statistics of Coal.
Holland's History of Fossil Fuel.
Parkinson's Organic Remains.
Johnson's Essays.
Dana's Muck Manual.
Field Book of Manures.
Report of Great Exhibition of Industry of all Nations, 1851.
Report of Juries of International Exhibition of 1862.
Patent Office Reports.
Geological and Agricultural Reports of Nova Scotia, Canada, and the several New England, Northern, Middle, and Western States.

Numerous miscellaneous articles from—

Silliman's American Journal of Science.
Farmer's Magazine.
London Mining Journal.
Penny Cyclopædia.
Hunt's Merchants' Magazine.
Mining Review.
Philosophical Transactions.
Scientific American.
London Engineer.
Proceedings of Geological Society, London.
Engineering.
Reports of American Institute, New York.
Edinburgh Philosophical Journal.
Annales des Mines.
Association Allemande, faits commerciaux.
Year Book of Facts.
Annual of Scientific Discovery.
Repertory of Arts.
Repertory of Patent Inventions.
Various Papers on Fuel, Iron, Gas, Steam, &c.

THE

UTILIZATION OF COAL DUST WITH PEAT,

FOR THE PRODUCTION OF AN EXCELLENT FUEL, AT MODERATE COST,

SPECIALLY ADAPTED FOR STEAM SERVICE.

FROM time to time, during many years, attempts have been made, both in Europe and in this country, to utilize various kinds of inexpensive or waste materials for fuel, the majority of which have had special reference to the fine coal and coal dust, technically denominated *slack*, which, it is well known, accumulates in immense quantities at the mines, and in no inconsiderable quantities on the wharves and in the coal yards of dealers and large consumers.

The amount of this material generally estimated to accumulate at the mouth of a mine is said to average full thirty per cent. of the entire quantity mined. This is considered not only as useless material, but an incumbrance upon the property and an item of expense in the business, requiring, as it does, no inconsiderable amount of labor to remove it, and area to store it, and so enormous have been these accumulations at some mines, in a series of years, that the owners have been obliged to purchase or secure "dumping ground" at a distance, and pay large sums annually for transportation in order to remove it and admit of the proper prosecution of mining operations.

This "slack" may ordinarily be said to be of the same quality as the coal of merchantable size, sent from the same mines, and consequently, of equal value, aside from the simple difference in size.

For some purposes, by means of furnaces specially adapted, it has been found practicable to make use of this slack in the condition in which it comes from the mines, and where this is practicable, it has a market value of about one third to one half the current value of merchantable coal. It is also used to a considerable extent in the manufacture of brick. The uses, however, to which it can be applied are limited, and the demand is of course limited, and at the present time is not sufficient to absorb the amount of screenings which inevitably accumulate in our city coal yards.

The first point generally sought to be attained in attempting to put this material into merchantable form, is to give it cohesion and size in such form as to render it available for use in much the same manner as the larger coals.

For this purpose it has generally been mixed with some adhesive substance of combustible character, as tar, rosin, asphaltum, or other resinous materials, formed into balls or blocks under pressure, and in this condition delivered for use.

It will readily be understood that a fuel compounded of such materials would be not only combustible but highly inflammable, and if perfectly made, would be of considerable value for certain purposes. It is, however, open to the objection, that thus far it has not been practicable to produce from such mixtures any other than an exceedingly sticky mass, disagreable to handle and liable to leave a black mark, not easily removed, on anything with which it comes in contact. The combustible character of the materials used is, moreover, so dissimilar, that the more inflammable portion, as tar, rosin, &c., is not only ingnited but almost entirely consumed long before the coal is burned out, so that unless great caution is used, the mass of fine coal will gradually fall through the grate bars or become imbedded in the bottom of the fire-box, so closely as to be almost impervious to a sufficient draft of air to sustain combustion, and is therefore self-destructive of the results sought to be obtained.

Added to this, is the fact, that in most of these cases the cost of the resinous materials, and the labor required, has been more than the real value of the fuel after it was in shape for use.

A method has been devised, and its practicability

thoroughly demonstrated and tested, by which this material may be rendered of the same value for fuel as merchantable coal, and that, too, by simple means, and at very moderate cost.

Numerous theories and plans have been proposed to accomplish this end, but without attaining, until now, the results essential to a business success in the enterprise, — the important feature being the production of a really *good fuel* at small cost.

For several years past, Dr. H. S. Lucas, of Chester, Mass., has devoted time and attention to the practical solution of this problem, and has, from time to time, obtained such results as gave promise of good success, until the spring of 1866, when his experiments, which until that time had been conducted on a limited scale, showed results of so marked a character that it was deemed advisable to test the whole matter on a somewhat larger scale, and in a more public manner; the results of which have given proof, beyond a question, of the perfect success of the method of manufacturing the fuel, and of its superior value as an article of fuel, especially for steam and smelting purposes.

The coal dust is mixed with crude peat in given proportions, and by a very simple process is manufactured into blocks of convenient size and form, which are dense and cohesive, and may be used for fuel very soon after passing from the mill, though it improves by exposure to the air for a few days.

The cost of manufacturing the fuel, after the materials are at the mill, is less than one dollar per ton.

The buildings and machinery necessary for producing it are quite inexpensive, and the laborers employed may, for the most part, be of the least expensive class.

The expense of transportation of the materials to the mill, and thence, after manufacture, to the place of sale or consumption, will generally prove to be the most important item f cost to be considered.

The most advantageous locality at which to make this fuel

will generally be found to be near to or by the side of a peat bog.

One very important discovery in relation to this matter is the fact that salt water, or marine peats, are equally valuable for this purpose as the best inland peats, and some of the experiments made, go far to prove that in certain respects they are superior for this purpose; so that, in this simple fact, we have opened up an immense value to marshes by the sea, which are almost without limit in extent, while at the same time they afford, for this special purpose, the material required, at the very point where the main item of cost (transportation of the coal dust), would be at its minimum; for vessels bringing it from point of shipment, can discharge directly at the place of manufacture, on or by the side of these marshes.

Our numerous lines of railroad, which now carry coal directly from the mines to various points inland, are, of course, equally available for transporting the same material in the form of dust or "slack," and many of them pass directly by or through extensive beds of peat, where it can be dumped and manufactured with the utmost facility.

For inland consumption, where supplies of coal are now received by vessels to the seaboard, and thence by rail, the same means are equally available for dust.

In all these cases, this advantage is also apparent, that the quantity required to be brought from the mines to yield a given amount of fuel here, is less, by a very considerable percentage, than of merchantable coal; that is, a concern which uses five thousand tons of coal would require to purchase only three thousand tons of "slack" instead, saving thereby, of course, the freight on two thousand tons. They also save the first cost of two thousand tons of coal, and the difference between the first cost of coal and slack on three thousand tons; against which is the cost of manufacturing the whole with peat, and the expense of re-shipment and transportation to the place of consumption or sale; the difference between these being very nearly the net gain or saving by the operation.

The following report of the use of this fuel on the Western Railroad is deemed sufficient, "without note or comment," to satisfy the most incredulous, that our statements in regard to its value for "*severe* steam service" are not exaggerated: —

CHESTER, MASS., Sept. 5, 1866.

T. H. LEAVITT, Esq.

DEAR SIR: I am happy to comply with your request to give you detailed statement concerning the trials made on the Western Railroad, from this point, with the compound fuel (peat and coal) manufactured under my direction, at the works of your company in Lexington.

The fuel arrived in good order, and, by the consent of the managers of the road, we were allowed to use the freight locomotive *Rhode Island*, built at Lowell in 1838, twenty-eight years ago, — weight, twenty-six tons; four feet six inch driving-wheel; sixteen inch cylinder; twenty inch stroke; two inch exhaust pipes; a wood-burner, one of the poorest on the road, and by no means adapted for burning our fuel to advantage.

The first trial was made 12th July last, on a regular trip from Chester to Washington, — the latter point being the "summit" of the road; and the section between here and there is well known to be the most difficult portion of the whole line to traverse, having several short and some double curves, with a grade of *eighty-three feet to the mile* for a part of the distance, and requiring the most severe steam service from locomotives.

The distance is twelve miles, and the total rise or elevation between the two stations is 950 feet.

We weighed and took on 1000 pounds of the fuel, and started from the station at 3.15 P. M., with sixty pounds steam; engineer, Theo. Dandurand, who has been on the road for ten years.

Our train consisted of eleven freight cars, three of them loaded, which is equal to fourteen empty cars, — a heavy train for this grade. Rail bad.

The grade, on leaving the station, rises for half a mile, and then descends for, perhaps, the same distance; and then commences the heavy grade.

Nine minutes after starting the steam had run up to 140, and we had to open the furnace door.

Twice we pumped cold water into the boiler, — once with both pumps, — when the steam fell ten degrees, from 130 to 120; but in five minutes was up again to 130. Had we been burning wood, and used both pumps in the same manner as in this case, the steam would have run down sixty degrees to 70, — so said the engineer.

The furnace door was open nearly four fifths of the time.

We made seven miles in thirty minutes, having passed the worst curves and the heaviest grade.

Here our fuel gave out, — steam standing at 130; and we were obliged to commence using wood for the remaining five miles of the trip. Steam soon fell to 120, and we were unable to raise it above that point.

We ran the first seven miles, by far the hardest portion of the route, in thirty minutes with our fuel; while the remaining five miles, with wood, took forty-five minutes.

The facts brought out are these: It will burn in any wood-burning engine, though, if it is to come into common use, we shall doubtless have fire-boxes specially adapted for it, which can easily be done; and a little experience will teach economy in its use. The exhaust pipes should be larger than for hard wood, perhaps three or three and a half inches. Combustion appeared to be almost perfect; there was no caking of the fuel in the fire-box, and it made but very little smoke. The heat is clear, steady, and extremely intense.

The engineer was astonished and delighted; said it was the greatest fuel for making steam that he ever used. He thinks the half ton, if burned in one of their large locomotives with six foot driving-wheel, would have carried a passenger train from Chester to Pittsfield, twenty-four miles. He thinks a ton of the fuel would take a passenger train over a common grade road, one hundred miles, and says a tender will carry four tons of it. If his estimates and opinions are correct, you will see at once that, at a cost of even ten dollars per ton for the fuel, it would cost but ten cents per mile to draw a passenger train; but, if the railroad company were to make it themselves, at your figures of first cost to produce, the expense would be but little over three cents per mile to run a train.

But, leaving figures and estimates for the present, the fact is fully demonstrated and established that this fuel will make more steam than any other of which I have any knowledge. About the same time we made two other trips with the same fuel. Of one I made notes, as in this case, and of the other I did not. They fully confirmed the results of the first trial, and were so similar in character that it is hardly necessary to repeat the story. With the improvements you have made in your machinery, and the apparatus you are now perfecting, especially adapted to the manufacture of this fuel, I cannot doubt that you will produce an article superior even to what you sent me, and, I should presume, at even less cost.

In order to a better understanding of the relative amount and cost of this fuel, as compared with wood, I should state that I went over the same route with the same engine, burning wood. We took on 2½ cords by measurement; were sixty-nine minutes running time between the stations; stood at Becket fifty minutes waiting for trains to pass; and, on arriving at Washington, found by measurement that we had consumed two cords lacking ten feet. Cost of wood for the trip, at $7.00 per cord, was $13.27.

Yours, truly,

HEMAN S. LUCAS.

CHESTER, MASS., Nov. 29, 1866.

I have read the foregoing statement, and hereby certify that it is in strict accordance with the facts in the case.

THEODORE DANDURAND, *Engineer.*

CHESTER, MASS., Nov. 29, 1866.

By invitation of Dr. Lucas, I went over the road on the engine, on two of the trips made with the mixed fuel, and am happy to give my testimony to the entire accuracy of the within statements. HENRY D. WILCOX.

HARTFORD, CONN., Dec. 10, 1866.

I was fireman on the locomotive "Rhode Island" at the time the trials were made. The above statements are strictly true. GEORGE MARSHALL.

The fuel used on these trials was manufactured at the works of the Boston Peat Company, at Lexington, Mass.

This company use and control the Leavitt Peat Machine, and their works are managed under the personal direction of Mr. Leavitt.

He has coöperated with Dr. Lucas in his experiments, and has produced a modification of his machinery, with additions, which are now perfected, and by means of which this fuel is manufactured at the rate of fifty tons per day, at a cost for labor of less than one dollar per ton. The machinery is simple, is easily set up and operated, and requires about ten-horse power to run it.

In the progress of these investigations and experiments, Dr. Lucas has likewise ascertained and developed the fact, that this improvement may be adapted, by certain modifications, to the smelting of various kinds of ores; also to the desulphurizing of ores, in the most perfect manner, — the process in all these cases involving but very small expense, compared with the cost of methods which now prevail.

The ores and flux are crushed, mixed with coal and peat in ascertained proportions of each, and the whole is manufactured together in the form of fuel, with which the furnace is charged at pleasure; and the results obtained show a superior product of metal at small cost.

The quick, intense heat generated, gives this fuel a decided advantage over any of the coals at present in use.

For the purposes of gas companies, this method of mixing and preparing fuels has some very important advantages.

It is well known that most companies use two or more kinds or qualities of coal for generating gas, — the poorer and cheapest kinds for volume, and the richer and more expensive, for body or strength.

Numerous experiments, some on a large scale, have clearly developed the fact that peat has a very considerable value for the same purpose, the volume being larger, while the strength or illuminating power is believed to be above the average of coals.

Doubtless still other uses of the principle of this invention will be discovered in time, but enough is already known to establish for it an important place in connection with the subject of cheap fuel.

APPENDIX.

We have felt a degree of pride at the exceedingly favorable mention which has been made of the former editions of this work by the press, in all directions, but we have experienced a much greater degree of satisfaction in observing the remarks, oftentimes at considerable length, which have been added, upon the subject-matter of which it treats, showing a quick appreciation of its importance, and a readiness, voluntarily, to aid in disseminating information concerning it, which is rarely accorded to any new enterprise.

Some of the following extracts are of general, others of local, interest; but all are worthy of careful perusal: certain it is, that all, with one accord, bear earnest testimony to the value of "peat as an article of fuel."

From "The New York Independent."

The Economical Value of Peat.—The great desideratum as to the use of peat has been to devise a method of consolidating it into a manageable and merchantable form at a moderate expense. The practicability of condensing it so as to produce an article of fuel of great value has been abundantly proved by numerous successful experiments. And it has also been shown by experiment that this fuel, when well prepared, has qualities which make it equal to any other, and, for some uses, superior to any.

To burn in an open grate, in a sitting-room, it is both economical and agreeable; and for an open fire in a sick-chamber (where none but an open fire ought ever to be allowed), it is invaluable as a purifier of the air, without the production of any sulphurous or other injurious gas. For steam purposes, it is quicker and more effective than any thing else, and would be of great use in driving our steam fire-engines. In making and refining iron, it is at least equal to charcoal; and, in the finer grades of iron-work, it is invaluable. It furnishes an illuminating gas, having, at least, double the power of coal gas; and many experiments place its comparative value much higher, while a great saving is effected in the manufacture by the entire absence of sulphur.

The one essential problem has been to invent an economical method by which its natural porousness could be overcome so as to solidify it in more convenient blocks, that can be handled and transported, and that will burn without crumbling to pieces in the fire. The Boston Peat Company, No. 49 Congress Street, Boston, have perfected a process which appears to be completely successful, and at such a moderate expense as to admit of general application wherever there are peat-bogs of the extent of even an acre or more.

By a machine of simple construction, but on a novel principle, the crude peat, wet as it comes from the ground, is dumped into a hopper, completely pulverized, so as to destroy the porousness, which no pressure can wholly overcome, and delivered in moulds, like bricks, which are dried in the open air for two or three weeks, and are then ready for delivery as good solid fuel, nearly as heavy as Liverpool coal. The machine costs fifteen hundred dollars, requires ten-horse power, and will work one hundred tons of crude peat in a day, making twenty-five tons or more of condensed fuel per day to each machine. The ordinary calculation in the city for that amount of steam-power is about twenty-five dollars a week; but as this machine would produce its own fuel, and may even be run with the refuse and unmerchantable stuff, the cost is much reduced. A further reduction would take place where several machines are run by one engine. The cost of ground for drying, and sheds for storage, may be less or more at pleasure. As neither the peat, nor the land it occupies, possesses any value in its natural state, all that is charged for royalty is simply so much wealth created by this great discovery.

There is no lack of peat in this country to supply abundance of fuel for centuries. It is found in all parts of the Northern States, and as far south as Virginia, in vast quantities. The Geological Report of the State of New York describes and locates more than ten thousand acres in the river counties alone. In the central and western sections, the quantities are almost illimitable. In Massachusetts, the Geological Report names nearly one hundred towns which have supplies of peat. Is is so in New Jersey, in Connecticut, on Long Island, and elsewhere.

In sections where other fuel is difficult to be obtained, and very costly, they may have by this process an abundance of the best and most pleasant kind of fuel, both for domestic uses and for manufactures.

When we consider that every acre of peat is calculated to be good for a thousand tons of condensed fuel, worth now eight or ten dollars a ton, at a cost of less than three dollars a ton for preparation, the amount of solid wealth which this invention will add to the country quite leaves petroleum in the background.

From "The Boston Journal."

The high price of coal, and the rapid destruction of our forests, ought to stimulate the people of New England, and indeed of the whole North, to an investigation of the feasibility of bringing *peat* into general use as an article of fuel. Wood is every year growing scarcer in New England, and the use of coal is fast spreading among country towns, where, a few years ago, the article was never seen. Our people have been prodigally wasteful of their wood-lots; and we are fast becoming dependent upon Pennsylvania and the British Provinces for all our fuel. As a consequence, we have been compelled to pay seventeen dollars per ton for coal in this city the past winter; and inland towns have paid eighteen or twenty dollars, while wood was sixteen dollars per cord. All the great cities on the Atlantic coast have been entirely at the mercy of coal operators and speculators, and most exorbitant prices have been asked.

Were wood and coal the only articles that can be used for cooking and heating purposes, we might be reconciled to this state of things. But there is in New England, and indeed throughout the North, a great abundance of peat.

One obstacle to the use of peat has been the difficulty of compressing it so as to render it less bulky, owing to its fibrous nature, which stoutly resists condensation. This difficulty has been overcome by machinery.

Experiments have recently been made, under the auspices of some enterprising merchants of Boston, for utilizing peat as a fuel. The experiments made here have been completely successful; and, if the sanguine expectations of those who have been engaged in this are realized, New England will, at no distant period, supply her own fuel.

It is well known that peat has been used for fuel, to some extent, in Eastern Massachusetts for several years. There is an abundance of it—thousands of millions of cords—scattered all over New England; and it is time, as we remarked at the outset, that our people were stirring themselves to see if it may not be brought into general use in order to stop the waste of our forests, now growing every year more valuable for timber, cheapen the cost of fuel, and render us less dependent upon coal monopolists and speculators for this indispensable article. It lies at our very doors, and there is no reason why its value should not be developed.

From "The Portland (Me.) Transcript."

Any thing that will give the people relief from the present high prices of coal, and tend to break up the monopoly of the coal-mining interest, while, at the same time, it saves our forests from further destruction, is worthy of a hearty patronage and encouragement; and we hope to see the attention of capitalists turned to the manufacture of peat. The people of Maine, so far removed from the coal-region, and paying so much for its transportation, should no longer neglect the rich deposits of fuel in the peat-beds scattered throughout the State. Dr. Jackson, in his Geological Survey of Maine, long since directed attention to the numerous valuable accumulations of this fossil fuel. He says,—

"The time may arrive when, even in Maine, wood becoming scarce, her neglected peat-bogs will be resorted to for fuel; though here, as in many other sections, were the *superiority* of the article over even wood or coal known and appreciated, the bogs would be worked now, rather than to await the period at which, for lack of other fuel, their valuable deposits shall be drawn upon."

He also says, that the localities of peat in Maine are so numerous, that it is hardly necessary to describe them, but points out localities in Bangor, Bluehill, Thomaston, Limerick, Waterford, and near Portland.

From "The Berkshire County Eagle."

The subject is one which should receive especial attention in New England, where the resources of fuel are not what they need to be.

Railroad men and manufacturers regard peat as superior to coal for the generation of steam, and the manufacture of iron and steel.

From "The New York Evening Post."

It is one of many unphilosophical things in the social history of our people, that so little progress has been made in the use of *peat*. It would seem as though we had been slow to believe that it was made to be used. As a natural product, under the wise ordering of things by the Creator, it is found in abundance in all parts of the country, as low down as the Dismal Swamp of Virginia, in positions easily accessible, and in forms and qualities suited for every purpose of fuel.

In its crude state, the peats most common are not so agreeable as wood or coal for domestic purposes; but the finer qualities of peat are preferable to either, and most of the common sorts are capable of being condensed into a perfect and most desirable fuel. Being entirely free from sulphurous and other objectionable ingredients found in coal, it makes a delightful fire in the grate, giving a charm to the parlor, and a delight to the sick-chamber. For raising steam, it is unrivalled by any other substance. One advantage is, that it creates no clinkers, and does not spoil the grate-bars like anthracite. In making the best iron, and in the finer processes of working iron and steel, it stands alongside of wood-charcoal. When charred, it is said to be better for welding purposes than charcoal itself. Many peats are also superior to the best bituminous coals for making gas. It is said that some varieties of peat are better for gunpowder than the charcoal made from dog-wood and alder.

In the preparation of peat for convenient handling and use, a great variety of methods has been tried. The oldest and best peats are so solid and fine, that they dry into a hard substance like cannel coal, without any artificial process except cutting the blocks of proper size out of the bog, and laying them on the ground to dry. But most kinds are too light and porous for convenient use, and burn with much smoke and less heat, unless artificially condensed. The most common method was by pressure; but this is expensive, and not fully effectual.

We have seen specimens of *condensed* peat, prepared by the Boston Peat Company, which appear to come nearer to the ideal fuel than anything else within our knowledge.

A single machine, driven by a six-horse steam power, will work fifty tons of raw peat in a day, which will yield about one fourth to one third of its weight in condensed fuel.

As a peat-swamp is generally unproductive, and therefore useless for agricultural purpose, it follows that the cost of fuel is the expense of manufacture, with the addition of the value which the demand may give to land otherwise worthless. The Boston Peat Company have patented their processes, and sell their machines.

From "The Syracuse Journal."

Peat as Fuel. — This article is soon destined to enter into lively competition with anthracite; and the probable effect will be that the "coal monopolists" will be brought to their senses and fair terms at the same time.

The article of peat as fuel was put to a test on the Central Railroad yesterday, and proved highly satisfactory.

There is an extensive bed of peat at Oswego Falls, opposite the village of Fulton, in Oswego County, on land owned by Bradford Kennedy, Esq., hardware merchant, of this city. For the purpose of testing its quality as fuel, a quantity was dried and prepared in the usual way, to be tested on one of the locomotives of the Central Railroad. Half a ton of that article, dried and ready for use, was sent down to this city, and Engine No. 106, a wood-burner, made ready for a short trial trip. Superintendent Lapham, of the Central, and Superintendent Van Vleck, of the Oswego Road, together with a party of other gentlemen, including Messrs. Howe and W. S. Nelson of Fulton, Derastes Kellogg of Skaneateles, Mr. Geddes, Jr., Mr. Southmeade of New York, and others, witnessed the trial. The locomotive was fired up at the Round House in this city, adjacent to the Central machine-shops, and run to Warner's Station and back, a distance of twenty miles, in the space of forty-five minutes. The engine drew but the car containing the excursionists, and was propelled at a moderate speed, without any attempt at "making time;" the object being merely to test the article of fuel. The peat made a beautiful fire, throwing out intense heat, and burned with a steady flame. The steam was kept at an even gauge of from ninety to one hundred pounds during the trial trip, and Superintendents Lapham and Van Vleck were highly pleased with the test of peat as fuel, pronouncing it a success. We understand that the usual amount of fuel consumed by coal-burning engines is a ton to every twenty miles; but, in this instance, only half a ton of peat was used, giving evidence of its value as a substitute for anthracite.

From "The Springfield (Mass.) Republican."

Peat contains the same chemical constituents as coal; indeed it seems to be only an imperfect form of the same material,—young coal, coal in the crude.

In Maine, beds of coal have been found in draining a bog, evidently formed from the wood of a species of fir, the balsam of which had been changed to bitumen, with which the deposit is very highly charged.

Peat has long been used for firing in some locations, especially in Ireland, where it makes a sweet and wholesome fire, safer for delicate lungs than either coal or wood. The heat is less drying, the ashes less troublesome, and the smoke does not irritate the eyes.

In its native beds, peat is heavily charged with water; and the want of a cheap method of drying and condensing it has prevented its being burned to any extent in this country; but modern researches have removed this objection, and means have been found for preparing it as fuel in large quantities, in merchantable shape, at a cost of four or five dollars a ton. The process has been patented; and the company, organized in Boston, has works now in operation in Lexington, Mass. The fuel is claimed to surpass coal for many purposes, especially for generating steam, and for the manufacture of iron and steel.

Peat-charcoal is denser than that from wood; and, as it contains no sulphur, iron made with it is of superior quality, and will not splinter.

Gas from peat has been used for some time in Paris: the hydrogen obtained is very richly carburetted, and is better than that from coal for illuminating purposes.

There are numerous indications that the stores of peat found in almost every township, the accumulations of past ages, will prove a rich inheritance to us and our children.

From "The Hartford Press."

The fuel question is already a serious one in manufacturing New England, which requires a greater supply for purposes of warmth, and for its thousands of engines, than almost any other part of the country. Our wood has long been scarce and costly; and we are so far from the great coal-beds as to make that material expensive. Attention has lately been directed to peat as a substitute for wood and coal, and, it is believed, with favorable results. A process has been discovered by which peat can be converted into a dry solid substance, in great quantities, at a moderate cost. The fuel so produced burns readily, gives a mellow but intense heat, is most agreeable for burning in the open grate, and is especially adapted to furnaces for generating steam.

From "The Madison (Wis.) State Journal."

The recent invention of machinery for pressing peat, in connection with the increasing scarcity of fuel in the West, and the exorbitant prices of wood and coal, has attracted attention anew to the extensive peat-beds near this city.

The subject of manufacturing fuel from this material was discussed quite prominently some ten years ago among our citizens; but the comparative cheapness of wood at that time, and the absence of any perfected machinery for reducing it to proper form, as well as the financial revulsion which shortly followed, led to a postponement of the project.

A further investigation of the qualities of this peat, of the extent of the deposit, and of the practicability of cheaply manufacturing from it an article of fuel equal in heating-power to the best coal, induces the belief that it will speedily become a source of immense profit to the fortunate proprietor, and of great advantage to the city and adjacent country. It will also, when worked into convenient compass for transportation, constitute an important article of export to the neighboring cities with which we have railroad connections.

We recently witnessed some experiments with it, as an article of fuel in a common wood-stove. The peat used was unpressed. The specimens burned with a flame clear and brilliant as seasoned maple or hickory, and produced no unpleasant odor like that of coal. From the trial we saw made of it, we conclude that it will make a most desirable article of fuel; and we trust the day is not distant when it will be made available, and this market supplied with it. If we are not mistaken, Colonel Slaughter, in opening this peat-bed, will, if he does not "strike oil," find a source of wealth not less valuable and remunerative.

From "The Madison (Wis.) Patriot."

We feel an interest, in common with our people, in keeping the subject of Wisconsin peat before the public, that it may so interest the attention of capitalists as to insure its more complete development, which we believe to be all that is necessary to bring it into general use, and prove it a source of very great benefit to our people. The middle, southern, and some of the western counties of our State, are but sparsely timbered; and the very limited quantity of timber is growing less every year, until the scarcity and price of fuel have become a question of serious inquiry with the people.

Peat or coal must be used as a substitute, — the latter we have not, nor can it be had, only at a heavy cost; but *peat* we have in abundance, and within our own limits. To this we must sooner or later resort, and the sooner the better for those who are compelled to pay the present high prices of wood.

The peat-beds near this city will, when developed, furnish an abundance of cheap fuel; and, as it is inexhaustible in quantity, its use will prove a source of wealth, not only to the enterprising proprietors, but to the country.

We have seen the peat tried, and find it an excellent substitute for coal. It has been used in our press-room furnace in driving our engine, and found equal to the best coal. We have therefore no hesitation in bearing witness to the good qualities of the peat, having used it; and unite most cordially in the general wish, that this rich mine of wealth may be speedily developed, and the peat brought within the reach of those who would so readily avail themselves of its use. Our wood lands are principally oak openings, and the wood is rapidly disappearing. The price is now from seven to eight dollars per cord. Dr. Hayes, of Boston, the first analytic chemist in our country, pronounces this peat equal to good oak-wood fuel, and for gas equal to the best cannel coal. Many experiments have been made with it in this neighborhood for fuel by our most judicious men, and their accounts correspond with Dr. Hayes's analysis. It must be of great value to the proprietors, as well as benefit to the country.

From "The Milwaukee Sentinel."

Upon a recent visit to Madison, we were shown by Mr. Hough, County Surveyor, a plat of the peat-bog lying six miles west of that town, and immediately upon the Madison and Milwaukee Railroad. In company with an intelligent Irishman, who informed us that, in density and endurance it was far superior to the Irish peat (not so inflammable, nor vested with the peculiar odor in burning), we instituted some experiments; in the first place by burning in a blacksmith's forge, where it gave out a steady, brilliant heat, though not as intense as that of bituminous coal, yet heating iron readily. We placed a peck of it with an equal amount of Briar Hill Coal in the open coal grate, and found it not only to outlast the coal, but to give a far preferable fire; quietly, pleasantly, not snappishly inclined, and free from the odious smoke and soot of coal, which will be a great *desideratum* to neat housekeepers.

The owners are making arrangements to work the beds extensively during the next season; and we predict for it a large sale, even within our own city, should its cost be even greater than coal, solely on account of its cleanliness. My credulity was heavily taxed while at Madison, through the stories told of its comparative value with wood in generating steam, at a steam saw-mill in the neighborhood of the beds; but I must confess myself astonished at my own experiments, proving it of far more value than I deemed possible, and worth not less than $1,000,000 to the fortunate owners, should no other extensive beds be discovered to mar its value. Surely such unheard of "diggings" ought to stimulate other explorations.

From "The Lewiston (Me.) Journal."

M. W. Farwell, Esq., has used peat in his house and under the boilers in his bleachery. He began his experiments last year, and, though the peat is not of the best quality, yet it proves to be so valuable that he will cut a thousand cords the present season. He regards it quite equal to charcoal. It makes a cheerful fire, and lights a room better than wood: its smoke does not irritate the eyes, nor does it obstruct respiration. It can also be used for many manufacturing purposes.

From "The Brunswick (Me.) Telegraph."

The introduction of peat is no mere fancy, but a subject of grave importance, especially in these times of exorbitant prices for coal, and unreasonable prices for wood. The "Journal" speaks of its making a cheerful fire. True; and one of our most cherished recollections is that of an old farmhouse in Byfield, Essex County, Mass., the residence of a grandfather. The kitchen fireplace was large enough to admit an ox-cart; and in that same fireplace always blazed in winter, a peat-fire, giving both light and heat. Beside it we have sat for hours, watching the roasting of potatoes, cracking of nuts, drinking of cider, and maliciously (little rascal that we were) eying the young rascals who were courting the girls, our respected aunts. Those were glorious times; and we have ever since had a fondness for peat as an article of fuel. The fires of love never burn dimly beside it.

From "The Newport (R. I.) News."

We have received from the publishers "Facts about Peat." The work is deserving of more than passing notice. No subject is of wider interest to the whole family of man than that of fuel; and its high price in this country for some time past, gives the subject a peculiar interest to us in America.

Hitherto the world has been dependent upon wood and the different kinds of mineral coal. It has not been a generally recognized fact that there exists another article, formed of wood-deposits like coal, but of a much more recent formation, and known by the name of *peat*, which is destined to be brought into an important competition with its two rivals, wood and coal. Many people in our own community, we will venture to say, never saw a specimen of it, and have scarcely heard of it.

Its supply is said to be abundant along the lines of our railroads, and in the vicinity of our machine-shops and founderies in all parts of the country where there is woodland. As the article of fuel is one of great expense in our domestic economy, this subject cannot fail to interest all; and the introduction of a new, cheap, and abundant article of fuel, to be dug out at our own doors, as it were, is an important matter. We commend this work, and the subject upon which it treats, to the attention of all.

From "The Fall River News."

We consider this a valuable work, especially at the present time, when coal is almost beyond the reach of the laboring classes. Hitherto we have been entirely dependent upon wood and coal, while almost every township has peat-beds sufficient to furnish the inhabitants with a fuel, in every respect, when properly prepared, equal, if not superior, to either. We cheerfully commend this book to the attention of all; for the whole community are interested in the subject on which it treats.

From "The Nantucket Inquirer."

The peat-beds on Nantucket and the adjacent islands are estimated at six hundred and fifty acres, from one to fourteen feet in thickness. The rapid destruction of our forests, and the constantly increasing price of wood, have now brought the article of peat to the notice of manufacturers and railroad companies, and will no doubt stimulate some enterprising Yankee to get up a machine that will press the peat dry, as it is taken from the beds, and turn it out in the shape and consistence of bricks. Then will our peat lands prove to be a mine of wealth to the owners.

From "The Biddeford (Me.) Journal."

The subject is beginning to excite considerable attention, not only on account of the fears of a growing scarcity of wood, its high price, and also that of coal, but because it forms one of the products of industry, which, when perfected from its raw state, forms, like mines of iron, lead, copper, and silver, great wealth to a nation.

From "The Brooklyn Union," Oct. 2, 1865.

PEAT.—A NEW PROCESS.—Through the benignity of the Creator, our country is richly provided with the means of counteracting the strikes of miners, combinations of dealers, and quarrels of corporations, which so often distract the coal market, and make fuel scarce and dear. In all directions there are to be found immense beds of peat, which is an excellent fuel for many purposes, even in its crude state. And we are assured that a very ingenious and simple process has been discovered, by which the chief inconveniences in the use of crude peat are removed, and a substance produced which is in some respects preferable to anthracite.

The process, which is patented, requires about two weeks to make the article fit for use, and then it is in as good shape for handling and tranportation as coke.

In this shape, it is found to be almost pure charcoal, easily lighted, burning with a clear fire, producing very little smoke, and leaving only a small residuum of ashes. The ashes are equal to those of wood for fertilizing purposes. The prepared peat has been used for raising steam, for wire-drawing, for brass-working, and for cooking and heating rooms; and for all these purposes it has been highly approved by good judges. If we are correctly informed, the company profess to be able to furnish fuel equivalent to a ton of anthracite for five dollars.

The development of such a source of wealth, and of general relief, lying all around in lands that are otherwise utterly valueless, will add at once to the general resources of the country, and to the means of comfort and life of all classes of society. If this new method is what it is represented to be, — of which we have satisfactory evidence, — we earnestly hope it may attract the attention of capitalists and business men without delay.

The high price of coal, the scarcity of wood, and the necessity for an abundant and consequently cheap article for fuel, have turned the attention of several of our own citizens to the peat which is found to a very considerable extent on Long Island. In other parts of the Eastern States, the interest of the people is already thoroughly aroused on the subject; and, if we are correctly informed, a company is established in Boston which is experimenting with peat, and has met with some very favorable results.

There is abundance of peat on Long Island, within reach of this city; and the season is not yet too far advanced for cutting and drying it for use in its natural shape. Perhaps there is now hardly time enough to get up establishments for its improved preparation this season, as some machinery is required to be set up. And yet, we believe, if the matter would be taken hold of *at once*, with a few thousand dollars of capital, and a moderate share of judgment and energy, something valuable might be accomplished even this year, and we should at least be in good readiness for another season.

From the same paper, Oct. 14, 1865.

MORE ABOUT PEAT. — The use of peat for fuel is but little known in this country; but it has become necessary to resort to it as a substitute for coal, as a remedy against strikes, extortion, and monopoly. If the experiments now in progress to consolidate the crude peat, so as to make it capable of being handled and transported without crumbling, and so that it will burn clear like coal, are successful, it will prove a formidable rival for anthracite itself. It is believed that deposits of peat suitable for fuel are scattered all over the Northern States in such abundance as would supply a very considerable part of the demand for fuel, and at a price much below that of anthracite, because it lies on the surface of the ground, is procured with very little labor, and, being found in almost every neighborhood, would make a great saving in the cost of transportation.

The great desideratum has been to contrive a process by which it can be put in merchantable shape at a cost not inconsistent with its value for fuel. Most of the numerous experiments, both in this country and in Europe, have resulted in nothing, either because they failed to effect the object, or that they were too expensive in working. A large part of the contrivances looked to the application of powerful pressure to solidify the peat into blocks convenient both for handling and for use; but powerful pressure is not only expensive: it does not effect the object. Peat is so porous and elastic, that it will not give up either its moisture or air by pressure.

A company in Boston have proceeded on a different and quite novel principle, and have at length completed the invention of a simple and rational process, by which crude peat, as it is taken from its bed, can be converted into a solid, dry fuel, in good shape, in large quantities, and at a moderate cost. The machinery required is simple, and not too expensive for use, and can be easily set up and run by the side of the peat-bed, wherever a small yard can be levelled for drying it in the open air. They are now prepared to furnish their machines at reasonable prices, with a guaranty that they will work as represented. The machine receives the crude peat as taken from the bog, and delivers it, in a very few minutes, *condensed*, and formed into blocks of any desired form, ready to be dried in the open air, and with but small cost for manual labor. Its tenacity for water is so far changed, that it dries in a small part of the time required for curing the ordinary peat.

We have seen specimens of the peat condensed by this process, and are acquainted with the principle on which the machine works, and think there is no reason to doubt its efficiency. The personal character of the principal managers is fitted to inspire confidence that they would not come before the public unless they had got a good thing, calculated to be a general benefit. Indeed, their method may fairly claim to be not only the best, but the only one, so far as is known, in this country, that is at once effectual, cheap, and rapid. They are the pioneers of the present movement in favor of the use of peat, having been engaged for several years in their experiments and inquiries. Their actuary, Mr. T. H. Leavitt, has published a pamphlet of a hundred and eighteen pages, containing more information — historical, scientific, and practical — about peat, we venture to say, than any other man in the country is possessed of. And his perseverance is well entitled to the success which seems about to be realized.

Those who wish to become thoroughly informed should procure this pamphlet; and those who would transact business with the company, should address the agents, Leavitt & Hunnewell, 49 Congress Street, Boston.

From "The Lowell Journal."

This is a good-looking pamphlet of a hundred and sixty-eight pages, giving us all the information upon the subject of peat that the most laborious and extensive research can possibly furnish.

The subject of fuel is one in which all of us are directly interested. Wood and coal, their different varieties and properties, are tolerably well understood; but peat, to a considerable extent, among us, is a new article; and even those who have used it in its crude state do not realize the extent to which it might be used, especially for manufacturing purposes, if properly prepared, and placed in the market.

From "The Prairie Farmer," Chicago, Ill.

This subject is attracting much attention at various points, on account of the scarcity and high price of fuel.

From "The Northern Farmer," Fond du Lac, Wis.

We have received from Messrs. Leavitt & Hunnewell, of Boston, Mass., a pamphlet on the preparation and uses of peat. With their improved method of preparing it, it bids fair to become of great value to our State, as we have, no doubt, abundance of it here. It may eventually fill up the greatest deficiency of our State, by furnishing an article equal to coal for fuel and smelting purposes.

From "The Kenosha (Wis.) Telegraph."

A Good Thing. — Madison, in this State, and Chicago, Ill., have been boasting for some time past of having in their immediate vicinity large beds of peat, which are capable of being turned to good account for fuel for domestic purposes, but more especially for mechanical purposes. Well, Kenosha cannot afford to be behindhand in any of these great natural resources; so she also boasts of her inexhaustible peat-beds.

Our fellow-citizen, Harvey Durkee, informs us that he has, on his farm, about two and a half miles from the city, a deposit of peat, which has been pronounced, by those qualified to judge, to be of the very best quality of that article. Old men who have used peat most of their lives, in Ireland, declare this to be in every respect equal to the best that country produces. This bed is three fourths of a mile in length, twenty rods wide, and ten feet in depth. Mr. Durkee has dried and tried some of this peat, and finds that it burns freely, makes a very hot fire, and leaves no residuum but a small amount of white ashes. He put a hodful of the prepared peat into his coal stove, and it burned as long, and gave out as much heat, as the best quality of hard lump coal. With the proper facilities for cutting and preparing the peat for use, we understand it can be furnished, probably, for $8 per ton; and one ton of the peat will go as far, and make as much heat, as two tons of the best Lehigh coal, for all mechanical purposes.

Now, if anybody or any company wish to start a woollen factory, or a cotton factory even, here is the material for the necessary fuel, so cheap as to throw the advantages of water-power into the shade. Since Mr. Durkee proved the qualities of this peat, we understand he is decidedly in favor of starting both a woollen and cotton factory, confident that no other locality in the State presents so many advantages for such manufactures.

From "The Brunswick (Me.) Telegraph."

Peat was first discovered in this town by Henry Putnam, Esq., about fifty years since, in the swamps east of Stetson Street. Several persons were much excited about it as a valuable discovery; but nobody was disposed to go into a peat speculation. I do not think, in the usual manner of cutting and drying, it can ever become a popular fuel. The difficulty of getting it thoroughly dry, would, in our uncertain climate, be a serious objection; but if it can be cheaply manufactured into neat, compact, solid blocks, I see no reason why it should not compete successfully with coal and wood.

This matter interests every one who has to buy a cord of wood. The article is abundant in this town: I presume all our swamps are underlaid with it. There is a large tract intersected by the McKeen Road, which drains into Mere Brook; the Duck Pond Swamp is another large deposit; the Dunning Swamp on Union and Pleasant Streets; another lies east of Federal Street; and another still farther east, extending from near the river, at the place formerly occupied by Mr. Bow, down to Humphrey's Mills; and there are doubtless many other localities in the town. N. S.

From "The Hingham Journal."

The use of peat as fuel is now attracting, generally, the attention of railroad men, manufacturers, and others; and, in this view, the issue of the pamphlet is timely. The treatise is prepared with care, and embodies much useful knowledge.

From "The Vermont Watchman."

FACTS ABOUT PEAT AS AN ARTICLE OF FUEL. — A well-timed and well-executed compilation of important facts. We never knew of any use of peat for fuel in Vermont, wood having been, and in some portions of the State still being, very abundant. There are regions in the State, however, where peat will be more economical than wood or coal; and we shall be glad to see it tried. Peat abounds in the State, from mountain tops to valley swamps; and doubtless in many places it is of sufficient depth and solidity for fuel. This book describes the article, the mode of cutting and curing, and its uses. For many purposes, good peat is at least equal to wood, and for some superior; while its use will drain swamps, and turn them to agricultural account, and prevent that denuding of the hills, of the woods, which is fast robbing us of water springs, and exposing the hills to be washed clean of the most valuable soil, and, in time, to become as ugly and sterile as the hills in the oldest parts of Maryland now are. Of all men, the farmers should turn their attention to peat, for the preservation of their best forests for more valuable uses than fuel.

From "The Pawtucket (R. I.) Gazette."

No doubt can be entertained by those who have even a limited knowledge of peat, that it can be advantageously used for fuel; but it has thus far received but very little attention in this country. Mr. Leavitt's facts and remarks throw a great amount of light upon the subject, and they ought to have a wide circulation. We have an abundance of peat; and the pamphlet before us tells us of its importance as an article of fuel, and how to prepare and use it.

From "The Lowell Citizen."

In late years, peat has come to be extensively used, not only for fuel, but as a source of motive-power, for the manufacture of gas, paper, gunpowder, and even for building and ornamental work. The questions of its supply, preparation, and most economical use, are of high interest; and this pamphlet embodies much needed information, which will aid in their solution.

From "The Essex County Mercury," Salem, Mass.

Peat has long been used to a considerable extent for fuel in different parts of Essex County; and not a few of the elder portion of our readers can well remember when it was much used in Salem. At the present prices of coal and wood, peat is much to be preferred to either of them, for most uses.

From "The Providence Daily Press."

"Facts about Peat" are not only interesting, but of the highest value. The company who have undertaken to develop the value of peat ought to be encouraged by all who have money to invest in new and probably remunerative channels of trade or manufactures. For ourselves we have often wondered how little use was made of peat.

The glowing heat and cheerful light of a peat-fire, are the very ultimata of a social evening; and our recollections of such a fire on the hearth at Johnny Campbell's, at the head of Loch Rannoch, are of the pleasantest character; so pleasant, indeed, that deprived of the sight of the fire, and the smell of the reek, we have even endeavored at times to find an insufficient consolation in the *taste* of the peat-reek in the genuine Glenlivit.

From "The Lawrence Sentinel."

There can be very little doubt of the utility of the great peat deposits in this Commonwealth, to which public attention has but to be rightly directed, to enhance greatly its value; and this work ("Facts about Peat") will be found to possess a permanent value.

From "The Waltham (Mass.) Sentinel."

The peat-bogs about us are represented to contain a large percentage of bituminous matter; and, when the peat is subjected to great pressure, it becomes a species of bituminous coal. Peat is said to be the last stage of vegetable matter before changing to coal.

If the working of the peat-bogs will save our woods, which are being cut down in such haste, and with so little of consideration, then we hope the attention of the people will soon be turned to this matter.

From "The Portland Advertiser."

We do not know how extensively this pamphlet has been circulated; but we are persuaded that its perusal by every consumer in New England would be productive of great good, and excite a new enterprise throughout the New England States.

From "The Middlesex Journal," Lowell, Mass.

Peat, as an article of fuel, has long been known; but it has not been so extensively used as its merits would warrant. The attention of intelligent men in manufacturing and railroad circles has, however, recently been turned to the subject; and we may hope, ere long, that it will take its proper place in the household, the manufactory, and on the railroad.

With coal at $16 and $17 per ton, the public have a deep interest in any thing which promises to render fuel more abundant and cheap.

In many localities in New England, and throughout the country, the earth is well stored with peat, which promises to add much to the wealth of those farms and districts of country where it is found.

From "The Boston Traveller."

Facts about Peat.— It is a thorough production; the author proceeding exhaustively, and arranging his abundant matter in a manner that renders the task of following him easy and profitable. He has mastered his subject, and evidently has neglected nothing that is calculated to illustrate it, and to press useful facts on the mind of the inquirer.

Various, minute, and copious in its facts, and showing how valuable is *peat as an article of fuel*, this work must have a great effect in directing attention to a neglected agent for the production of heat; one which Providence has placed most freely at the command of man, and which ought to be made to enter very largely into human consumption.

Mr. Leavitt is literally correct when he says, that the substance of which he treats so well "is of sufficient importance to command earnest attention, not only from the business man, on the score of its application to domestic purposes, manufactures, and the arts, but from the philanthropist, in view of the relief it may be made to afford as one of the necessaries of life."

It needs only that the value of peat should be understood to bring it into general use, to the great relief of all interests.

From "The Scientific American."

In peat we shall find an economical substitute for coal at its present prices (April, 1865), and even at rates much below; for the marketing of the former substance, or preparation of it so as to render it available, must certainly cost far less than coal.

No shafts have to be sunk, no extensive and costly system of engineering and surveying are needed; and beyond the expense of the machinery for *condensing* it, little seems to be required to utilize the deposit with which Nature has covered large tracts of land in this country.

The testimony of scientific men is freely given as to its value.

From "The Syracuse Journal."

Dr. B—— is making arrangements to prepare his peat for market. Now let us estimate the quantity upon the fifteen acres. Fifteen acres, at an average depth of eight feet, will produce 40,836 cords. Estimating a cord of peat to be worth a cord and a half of hard wood, there will be equal in value to 54,448 cords of hard wood. Estimating wood at six dollars a cord, the total value of this peat-bed, when marketed, will be $326,688. Allowing sixty-six and two thirds per cent. cost for preparing and transporting it to market (which is a large allowance), and there will remain a net profit of $108,896.

Peat emits a considerable flame — about between hard coal and hard wood; it leaves no cinders to sift; it burns equally well in a coal-stove, wood-stove, or fireplace, and makes a very pleasant fire.

From "The New York Evening Post."

The high prices of coal are having an effect which will soon be turned to the advantage of the community. Already they have brought into existence several enterprises, which, in a few months, will produce a large supply of fuel from the *peat* deposits of this country; and it will be likely to come into close competition with the fuel from the coal mines.

The probability of continued high prices, together with the favorable results of recent experiments with *peat*, and new discoveries of it in quantities, have called the attention of many business men to this substance, as a new source of supply.

The burning or heating properties of the best peat are nearly equal to those of anthracite coal.

There are many peat-beds in this State. A trial of it is making at the American Institute Fair against coal, with satisfactory results.

Great interest is felt by many citizens to whose knowledge these facts have come, and they are confident that a material change will eventually take place in regard to the fuel supply of the country. The tendency will be, in any event, to protect the public from speculations and monopolies in coal.

From "The American Artisan."

The very high price which coal has lately reached in this country has led some enterprising capitalists to turn their attention to the subject of obtaining supplies of peat for fuel.

There are in the United States, large quantities of this valuable material. In Western New York and on Long Island there are extensive beds; and there is little doubt that in and near the neighborhoods where it is found it may be made to serve as an economical substitute for coal, even when that is at a much lower price than at present (August, 1864).

From "The Lockport (N. Y.) Daily Union."

The rapid advance in the price of wood and coal within the past few years, and the near approach of the time when wood, on account of its scarcity, shall cease to be generally used as fuel, have led many to investigate the practicability of bringing into use as a substitute, peat, which was known to be in large quantities in various parts of the country. It was found to be impracticable to be used as fuel in its natural state, on account of the foreign matter, and its unwieldiness and bulk; being thus impossible to make it a mercantile commodity, and difficult and unhandy to use. Many experiments have been tried by which the foreign matter could be separated, and the peat put in some neat and compact form, thus making it easy to handle. More or less success has attended the various attempts; and in some parts of the country it is rapidly being brought into use.

In this country, we have inexhaustible beds of peat; and if this experiment is the success that is hoped, we need no longer tremble as we see our forests rapidly falling away, or sigh when we read of a strike in a coal mine. It is estimated that the cost of manufacturing will be one dollar per ton. All that it will bring over three dollars will go to the manufacturer as a profit and interest on the capital invested in the bed. Prices, however, regulate themselves in accordance with the demand; but it is hardly probable, with all the peat-beds in the country producing fuel, that wood and coal can lord it much longer in their present manner. One ton of this peat is estimated to burn as long as a ton of coal or two cords of wood. The smoke from it is much like that from hickory wood — thin and blue; there appeared to be no unpleasant odor; and the ashes are not troublesome, like coal-ashes. In fact, there appears to be no reason why, if it can be put in merchantable form, that it should not become our staple article of fuel.

The prospect of such a revolution in fuel will undoubtedly interest the community at large.

From "The Springfield (Mass.) Republican."

The high price of coal and wood is very naturally turning the attention of the public to the vast beds of *peat* which exist in various parts of the country, with the hope of finding in them the much-desired cheaper fuel. In old countries, peat has long been used as an article of fuel, especially among the poorer classes, who have, in fact, known nothing else. But in this comparatively new country, where hitherto both wood and coal have been abundant and cheap, our peat-beds have, for the most part, been allowed to lie unmolested, as the product could not be taken out and prepared for market at a cost low enough to make it any object to bring it into competition with other articles of fuel.

But with ingenuity, stimulated by the present high prices, the problem of how to prepare peat for market at a reasonable cost bids fair to be speedily solved, if, indeed, it is not already satisfactorily answered. Of the value of peat, properly prepared, as an article of fuel, there is no question. Besides its worth for domestic purposes, in which the majority of people are most interested, it is unrivalled by any other substance for raising steam, and has lately been tried with marked success in the locomotives of the New York Central Railroad. Being free from sulphur, peat is also well adapted for the reduction of ores; and in making the best iron, and in the finer processes of making iron and steel, it is equal to wood-charcoal; and, when charred, it is pronounced better for welding purposes than charcoal itself; while some kinds of peat are equal to the best bituminous coal for making gas.

At present, the great bulk of peat would make the item of freight a large one if it was transported any great distance. But we do not see why, already, peat cannot be furnished to those who live near the beds, at a cost much less than coal, to which it is said to be equal, ton for ton, for many purposes. Labor and freight are, of course, the principal items in the cost of peat; and, as soon as quick and cheap methods of extraction and condensation are really discovered, there will be a large use of peat, and some of the profits that now go into the pockets of owners of coal mines will be transferred to the proprietors of peat-beds.

It may not be generally known that there are almost inexhaustible peat deposits within a few miles of this city. Mr. Reuben Brooks, of West Springfield, owns a large peat-bog, and there are others in West Springfield and other towns in this vicinity. Mr. Brooks has long used peat in his own family with satisfactory results, burning it in a common coal grate, where it gives forth a blaze like wood, and a heat much softer and pleasanter than that from the common anthracite or even bituminous coals. We have no means of knowing how the price of it delivered would compare with the price of other kinds of fuel; but peat is going to be the fashion before many years, and Springfield is to be congratulated on having a large supply close at hand.

From a Paper recently read before the British Association, by D. K. Clark, C. E., of London.

"TORBITE:" A NEW PREPARATION OF PEAT.—The writer had occasion a short-time since to inspect professionally the works established at Horwich, in Lancashire, to manufacture fuel and charcoal from peat, and was so struck with all that came under his notice, and impressed with the importance of the results obtained, that he feels that he cannot bring a more interesting subject before the meeting.

The question of the manufacture of peat into fuel is in reality a question of supplementing the natural supplies of coal with a fuel which may be made superior to it in every respect, more abundant, and more readily accessible. The consumption of coal is so enormous, and goes on annually increasing at such a rate, that, for some time past serious apprehensions have been entertained that our coal measures will be exhausted at no very distant period. Our stock of coal, excluding all that lies at a greater depth than 4000 feet, has been estimated at 83,544,000,000 tons. In 1863, the consumption reached 86,300,000 tons; and the average rate of increase for the last ten years has been two millions of tons a year. Thus, supposing our stock to have been correctly estimated, in less than a hundred years our coal will be exhausted. Fortunately, however, Nature has not left us dependent on our coal measures alone, but has also given us our bogs.

Peat, it is well known, possesses many most valuable properties as a raw material for fuel; but the attempts hitherto made to utilize peat on a large scale have proved failures, owing to the difficulty of dealing with a substance exceedingly bulky, very loose, and holding from seventy-five to eighty-five per cent. of water.

To separate the water, and to condense and mould the peat into convenient sizes at a cost sufficiently low to render it commercially available as fuel, is a problem which has baffled the efforts of many operators. In most instances, compression has been applied for the purpose of imparting the requisite degree of solidity, by means of powerful hydraulic presses, or other machinery. In the process adopted by Messrs. Gwynne and Mr. C. Hodgson, the peat is first dried and powdered, and then pressed into blocks; but the action of compression is purely mechanical, and, though it imparts great compactness by bringing the particles of the peat into close contiguity, *it does not really solidify the substance*, since, on being exposed to heat, *it resumes its original form, and crumbles to pieces.* Fuel thus prepared is totally incapable of resisting the action of a blast, or even of a moderate draft; and, though Mr. Hodgson still carries on the manufacture of fuel by this process, the consumption is very limited.

According to Mr. Cobbold's mode of treatment, the peat is immersed in water for the purpose of separating the fibre from the more decomposed matter, and the water is afterwards got rid of either by simple evaporation, or by means of centrifugal power; but, though by this means a very dense fuel is obtained, the separation of the fibre deprives the fuel of coherency, besides which the process is laborious and costly. Attempts have also been made in Ireland to utilize peat by manufacturing it solely for the sake of its chemical products. Many valuable products have thus been obtained, from which even paraffine candles have been made; but the cost far exceeded the market value.

But such attempts have not been altogether in vain, inasmuch as the experience thus gained in the treatment of peat has proved of great value. To know what will not do, is a great step towards knowing what will do. By the system of manufacture at Horwich, mechanical compression in any manner is studiously avoided, being not only costly, but also ineffectual. The means of separating the water suspended in the peat have been carefully perfected. The necessity of dealing with and getting rid of such a large proportion of water has been a standing difficulty from the first, and the cause of excessive expenditure. At Horwich, the problem has been carefully studied, and the difficulties appear to have been successfully overcome. Until a mode of artificially drying peat rapidly and economically had been worked out, air-drying was necessarily resorted to; and where limited quantities of fuel — say about a hundred tons a year — only, are required to be made, air-drying may suffice; but for large quantities it would be, in our fickle climate, too uncertain a process to be depended on, and for seven months in the year, it would not be available at all.

According to the system matured and established at Horwich, the peat, as it comes from the bog, is thrown into a mill expressly constructed, by which it is reduced to a homogeneous pulpy consistency. The pulp is conveyed, by means of an endless band, to the moulding machine, in which, while it travels, it is formed into a slab, and cut into blocks of any required size. The blocks are delivered by a self-acting process on a band, which conveys them into the drying-chamber, through which they travel forwards and backwards on a series of endless bands at a fixed rate of speed, exposed all the time to the action of a current of heated air. The travelling bands are so arranged that the blocks of peat are delivered from one to the other consecutively, and are by the same movement turned over in order to expose fresh surfaces at regular intervals to the action of the drying currents; so that they emerge from the chamber dry, hard, and dense. To the peat substance thus treated the name of "torbite" has been given, from the Latin *torbo*, by which name peat is constantly mentioned in ancient charters.

The next stage in the process is the treatment of the torbite in close ovens, when it may either be converted into charcoal for smelting purposes, or may be only partially charred for use as fuel for generating steam, or in the puddling furnace.

The whole of the Horwich system has been planned with a view to the utmost economy of time and labor. The raw peat is nearly altogether automatically treated by steam power. Introduced at one end, it issues from the other in the form of charcoal within twenty-four hours after it is excavated from the bog; and the manual labor expended is almost entirely limited to the first operation of digging: consequently the actual outlay in labor and fuel, in the production of the charcoal, does not exceed from 10s. to 12s. per ton; but, in addition to the economy thus effected by charring in close ovens, a considerable quantity of valuable chemical products are yielded, as ammonia, acetic acid, pyroxylic spirit, and paraffine oils, the sale of which alone will nearly cover the expenses of the whole process.

The fatty matter, separated by distillation, forms an excellent lubricating grease; the yield of which averages about five per cent. of the weight of charcoal produced. In its crude state, it has been sold for £12 per ton at Horwich.

The charcoal made from torbite is extremely dense and pure. Its heating and resisting powers have been amply and severely tested, and with the most satisfactory results. At the Horwich Works, pig-iron has been readily melted in a cupola. About 80 tons of superior iron have been made with it in a small blast furnace, measuring only 6 feet in the boshes, and about 26 feet high. The ore smelted was partly red hematite, and partly Staffordshire; and the quantity of charcoal consumed was 1 ton 11 cwt. to the ton of iron made; but, in a larger and better constructed furnace, considerably less charcoal will be required. It has also been tried in puddling and air furnaces, with equally good results, considerably improving the quality of the iron melted. For this purpose, the fuel was only partially charred, in order not to deprive it of its flame, which is considerably longer than that from coal. Some of the pig-iron made at Horwich was then converted into bars, which were afterwards bent completely double, when cold, without exhibiting a single flaw. Messrs. Brown & Lennox, in testing this iron for chain cables, have reported that its strength was proved to be considerably above the average strength of the best brands.

In Germany, peat mixed with wood-charcoal is very extensively used in the production of iron; the peat, as prepared there, not being sufficiently solid to do the work alone; but it is found that the greater the proportion of peat that can be used, the better is the quality of the iron produced. The gas delivered from the high furnaces has also been satisfactorily employed in the refining of iron and the puddling of steel. The value of peat in the production of iron has long been established. Iron metallurgists are agreed in the opinion that iron so produced is of very superior quality. In every stage of iron manufacture, and in welding, peat-charcoal is most valuable. At Messrs. Hick & Son's forge, in Bolton, a large mass of iron, about 10 inch square, was heated to a welding heat with peat-charcoal, made at Horwich. The time occupied was less than the operation would have taken with coal; the whole mass was equally heated through without the slightest trace of burning on the outside; and in hammering out the mass, as much was done with one heating as ordinarily required two heatings to effect. The importance of obtaining an abundant supply of peat-charcoal, at cheap rates, cannot, therefore, be too highly estimated.

For the generating of steam, the fuel made at Horwich has also been well tested, and its superiority over coal, practically demonstrated, both in locomotives and stationary engines. On the Northern Counties Railway of Ireland, a train was driven with it from Belfast to Port-rush, a distance of seventy miles. The result at the end of the journey showed a saving, as regards weight consumed, of 25 to 30 per cent. over the average of three months' working with coal on the same journey. There was an excess of steam throughout the run, though the fire-door was constantly open, and the damper down. At starting, the pressure was 100 lbs., but during the trip, and while ascending a steep incline, it rose to 110 lbs., and afterwards to 120 lbs., with the fire-door open. While running, there was no smoke, and very little when standing still.

At the Horwich Works, the fuel was tested against coal under the boiler there. This was done on two consecutive days, the fire having, on each occasion, been raked out the night previous. The following results were obtained: Coal got up steam to 10 lbs. pressure in two hours 25 minutes, and to 25 lbs. pressure in 3 hours; peat-fuel got up steam to 10 lbs. in 1 hour 10 minutes, and to 25 lbs. in 1 hour 32 minutes; 21 cwt. of coal maintained steam at 30 lbs. pressure, for 9¾ hours; 11¼ cwt. of peat-fuel maintained steam at the same pressure for 8 hours. But, in addition to this, a large economy is effected by the use of peat-fuel for the generation of steam, in the saving of boilers and fire-bars from the destruction caused by the sulphur in coal, from which peat is free. In Bavaria, peat-fuel has been used on the railways for several years past; and the economy effected by its use in the wear and tear of the engines is stated by the officials, in their reports, to be very considerable.

The bogs of Great Britain and Ireland cover an area exceeding five millions of acres, the average depth of which may be taken at twenty feet. Nature has thus supplied us with the means of adding to our stock of fuel some twenty thousand millions of tons. In Ireland, about a million and a half of acres have been thoroughly surveyed. In the reports of these surveys, it is stated, that beneath the peat an excellent soil, well situated for drainage, was found, fit for arable or pasture land.

When it is considered what peat is capable of doing, and all the results involved in the question of utilizing peat, it is impossible not to feel impressed with the conviction, that the foundation has been laid of an undertaking of great national importance and interest.

From "The American Railway Times."

STEAM FUEL.—Wherever the main source of artificial motion may lie hidden away, awaiting ultimate development, whether in air, or in water, or in the heat of the earth itself, matters less to the practical man than to the philosopher.

Coal has long been the main source from which that power has been obtained. How much longer it may continue so to be is uncertain, not so much from any immediate probability of failure in the supply, as that, of late, other substances have been utilized which hitherto were comparatively unknown, or considered inapplicable to the purpose of steam generation. It becomes us to consider, not only how to economize that supply, which we can at present call our own, but how to produce a fuel which shall satisfactorily occupy its position as well now as when our coal-fields cease to yield.

The two main sources from which the present generation may expect to derive practical benefit, and to which we may look for aid in the economizing of our coal, are *peat* and *petroleum*.

The deposits of peat in Great Britain and Ireland occupy an area of about six million acres. The thickness of peat varies in different localities, from two to forty or fifty feet. Assuming the average thickness to be only twelve feet, an acre would yield about 3500 tons of dried peat; consequently the aggregate estimated acreage in this country would produce twenty-one thousand million tons of dried peat, equal to the supply of twenty-one million tons per annum for a thousand years. It cannot be supposed that these enormous masses of vegetable matter were created to be either useless or noxious. Nor is it a matter for wonder that attention has often been directed in this country, and in others where similar deposits exist, to the means of utilizing the peat, and reclaiming the land which it covers.

The value of *peat*, when properly dried, is well known and admitted, both for domestic fuel and for generating steam; and charcoal properly made from such peat is, in all respects, equal, if not superior, to wood-charcoal. When dug from the bog, peat generally contains from fifty to seventy-five per cent. of water.

The inference drawn from practical experience is, that, to insure commercial success in utilizing peat, the operation must be inexpensive and expeditious, costly machinery being avoided.

From four to five tons of peat, as taken from the bog, are required to make one ton of dry condensed peat. The cost varies in different localities; but it may be safely assumed that the average cost will not exceed that of coal at the pit's mouth. Peat thus prepared burns very freely, will stand a powerful blast, emits great heat, is smokeless, and produces less ash than the average of coal or coke. It is impervious to water, improves by keeping, and is incapable of self-ignition. From two and a half to three tons of prepared peat will make one ton of excellent charcoal, according to the degree of carbonization required.

The general heating power of the condensed peat has been proved to be very superior to that of coal; and, in fact, this article appears to be well adapted as a fuel for steam-engines, whether marine, stationary, or locomotive. Its use has been found to effect a saving of fifty per cent. in time in generating steam, and it will do double duty as compared with coal. The absence of smoke and clinkers, and the preservation of furnace-bars and boilers from the destructive effects of sulphur from coal, are additional and important advantages.

The locomotive engineers of three railways in Ireland united to carry out a practical trial of the condensed peat on the Belfast and Northern Counties Railway, with the view of testing its qualities as a fuel for locomotives. The trip was made from Carrick Junction to Ballymena, a distance of twenty-seven miles. During the whole of the journey there was an excess of steam, although the fire-door was kept continually open, and the damper down, for the greater portion of the distance. The pressure at starting was a hundred pounds per square inch. The commencement of the journey was up an incline of one in eighty, four miles long, and with double curves; while ascending the incline, the pressure rose to a hundred and ten pounds, and afterwards to a hundred and twenty pounds; and this with the fire-door open. The speed was about forty miles an hour. While on the way, the fuel emitted no smoke, and very little when at stations. The fire-box was examined at Ballymena, and a very small portion of clinker was found. The smoke-box was perfectly free from cinders or dust, — a proof that the fuel had stood the blast exceedingly well; and it is the recorded opinion of the experimenters that the condensed peat was in every respect well adapted as a fuel for locomotive purposes.

In the face of such results as these, bearing the testimony they do to the fitness of properly prepared peat as a steam-fuel, the wonder is that it has not been generally brought into use. One reason why it has not, may lie in the limited quantity manufactured for steam purposes, the greater value of peat lying at present in its conversion into charcoal for smelting, for which purpose it is used in considerable quantities with the best results. Another cause for its non-adoption may be the hesitancy to depart from the old beaten track, which so often stops the way of improvement. The success of the practical trials it has undergone ought to be sufficient to commend its further use. No serious alterations to machinery are involved in its adoption, the only thing necessary being a reduction of space between the fire-bars to insure perfect combustion.

Such a substitute for coal or coke deserves attention. The comparative absence of smoke, and the total absence of all sulphurous vapors, ought to be a sufficient inducement, independently of the economy effected.

The question of the use of peat in locomotives is not a new one. About twenty years since Lord Willoughby d'Fresby had some tried in the Hesperus Locomotive on the Great Western Railway. This engine was of Hawthorn's patent return-tube construction, and required about one third more peat than coke, with equal drafts. Mr. Vignoles has also interested himself in the same direction. Opposite opinions, however, have always existed as to the economical merits of peat; but it may yet prove an efficient substitute for coal in all its varied uses.

From an Article in the New York Reformer, by Dr. W. V. V. Rosa, of Watertown, N. Y.

What is peat? Where should we look for it, and how shall we know it? In what is it better than wood or coal, and in what respects inferior? Is it scarce or abundant? Is a peat-bed of much value in money? and what is the history of peat?

All these, being questions of keeping warm and cooking and gas and motive-power and making money, are asked almost daily; for peat is becoming a subject of active interest everywhere at present from the high prices we have to pay for wood and coal.

All over the State, just now, solitary individuals in high boots, and trousers tucked in, with a long pole over the shoulder, and speculation in their eyes, may be seen following a sloppy, crane-like course of life, mysteriously wading about where bullfrogs most do congregate, who, if they are asked what in the world they are doing way out there, return a very swampy kind of answer; while they hurry on and leave you as clear as mud upon the mystery of their queer accoutrements, and unaccustomed advent, and their marshy ways. Yet, if you become familiar with the matter but just a little, the fog will rise from the subject, and you'll understand it easily, that it is *peat-beds* they are looking after; and that, if one should happen to exist upon your farm, it may be an excellent thing for yourself to know, as an acre of peat may be worth a thousand dollars, or two thousand even, instead of being a worthless bog and a nuisance, so given up to croakings and paludal dirges, that even abundant liquidation cannot still these "voices of the night."

But to return to the peat, — to the what and where is it. In answering, let us review for a moment what we all know concerning wood. It will assist in following the changes which take place in one form of it, — the vegetable fibre of mosses and ferns, for instance, — while it is passing into peat; most peat being the product of partially decomposed and partially preserved beds of mosses and ferns in swampy places.

Wood, then, is a compound substance, mainly carbon, — that is, coal — united variously with some mineral substances, such as potash, lime, silex, together with gases, — oxygen and hydrogen, and with water, &c., in the form of gums, resins, starch, sugar, and the like, in great variety.

These substances, in burning, form new compounds, such as carbonic acid, creosote, wood-vinegar, naphtha, alcohol, and the like, which pass away in smoke and vapor; and the other parts remain as ashes.

If, however, we wish to convert the wood into charcoal, the process is controlled and modified somewhat. The wood, gathered in bulk, is covered over thickly with earth, to *prevent free* access of air; a very little being admitted below, sufficient, and a trifle more, however, to consume that portion near the air-holes. A tolerably high heat is thus diffused throughout the pit; and the slight excess of air thereby quickened in its action, soon causes new combinations to take place, and decomposes and carries off the more destructible parts; and then, the draft being closed, the fire goes out, the pit cools down, and, the earth being removed, the coal is ready for use. By this means, excepting just about the air-holes, only a part, the more volatile and destructible constituents of the wood, are burned or recomposed, or pass away, while the main part — the carbon and mineral part — is left unconsumed.

Now what is understood by burning? When we say a substance burns, it signifies usually that the substance — coal or gas, wood, weeds, grass or moss, for instance — unites *very rapidiy* with oxygen, which is abundant in the air; the substance burned being thus changed in its form, but not destroyed, not annihilated, as that, of course, would be impossible. All that existed in the wood before, still continues to exist, though in other shapes; mainly in gases, partly in liquids, and partly in mineral, as in the ashes. During this process, or rather by it, heat is created. If the process goes on very fast, it is very hot; if slow, less hot; and though so very slow indeed that no heat can be perceived, the burning is in reality still going on, though the heat is unobserved. It will exist, however, though in a degree too slight to cause sensation of the slightest warmth.

Metals — iron, for instance — may burn, that is, unite, the same as wood or coal, with oxygen. In this case but little gas is formed; nearly the whole remaining as oxide of iron; that is, as rust. Rust is the ashes of iron. If the oxidation is rapid, as when it is burned in a jar of pure oxygen, great heat and light are caused; if slow, as when iron rusts in damp air, or under water, none is observed; but yet the rusting of iron under water is as really a burning of the iron as when the same occurs in oxygen, or at the forge, with intensest heat and light.

Water, indeed, being composed in part of oxygen, and holding a little extra in solution, is a good substance (strange as it may sound) to burn things with; in some instances better by far than air. And this is an essential point to observe in studying the formation of peat, that water is a good substance to burn things with; that is, if you are in no hurry, if you have years to spend in burning a very little, — so very slowly that an insurance policy might run out and be renewed, and out again a score of times, before the job is finished.

Although water, by preventing the free contact between actively burning bodies and the air, will "put out" fire, that is, will stop the *rapid* combustion which air favors and supports, still in reality the water does not put out the fire absolutely, as chemists would define the term, but rather, in many cases, makes its continuance certain, though centuries might be the measure of the slowness of the work.

One thing more, essential: Though water will burn many things better than air (even iron for instance, which, unless very highly heated from without, will not rust at all, that is, not burn, in *dry* air), still it is slower, or will *not* burn other substances which are easily consumed in air. And *coal* is just one of these. Coal will not oxidize, that is, not burn or rust or decay in even hot water. It will keep there forever.

And now, it being clear that water prevents *rapid* oxidation by excluding *free* access of air, and yet insures its slow continuance to a certain stage by furnishing a little, and that it burns and converts most substances easily or surely, and stops at others, among which is coal, it may be understood with but little further thought, how peat is formed, and where it is most likely to be found.

We can see that where large quantities of woody substances, such as mosses, ferns, &c., are for a long time accumulated, and remain always thoroughly soaked with water, as in many swampy localities, such places may be considered much like very slow-burning coal-pits; that they are places where mainly, by an exceedingly lingering oxidation, new compounds and recompositions take place, and the more *easily consumable* portion of the vegetable matters there gathered, being volatilized or burned, pass off, and leave the coaly portion especially, unconsumed, much in the same way, in principle, that it is made and left in ordinary pits, the water acting here, in part, as the earth covering does there, to govern and moderate the changes and oxidation by preventing free access of air, yet allowing or furnishing a little; and finally, when the coal-state is reached, the bed being already cool, the oxidation, absorptions, and recompositions cease, and the carbon, ready for use, is preserved for centuries.

When wood or vegetable fibre dies, and remains in places freely exposed to air and sun, it is soon almost wholly decomposed, passing away in gases mainly, as has been mentioned; and but very little of it remains. If, however, in a *cool* climate, and other circumstances being favorable, it falls in large amount into places kept always thoroughly wet, then the decomposition is only partial, and the most of the carbon remains.

A peat-bed, therefore, during its formation certainly, must be looked for, at least in this climate, in swampy, boggy places, yet not too wet for vegetation to exist.

The great body of most peat-beds result from a variety of moss called *Sphagnum palustre*; in part also from ferns or brakes, rushes, reeds, and other plants; and even fallen timber occasionally constitutes a part. This bog-moss, of which there are many varieties, flourishes luxuriantly in such positions; its fibres growing even a foot and a half in length. And it has this quite remarkable peculiarity, that, while its old roots and lower portions die in the wet soil or water, new roots spring out from the stem above; and thus, without interruption, it continues to grow and accumulate year after year, in a deep and deeper bed for centuries, at length resulting in a vast collection of carbon, frequently almost as pure as bituminous coal.

Peat-beds, however, are not always wet places after their formation, though they must have been so during it. The swale may have been gradually filled up with the growth of peat itself, and then soil may have been formed or washed over it. In some instances, — as over parts of the Oswego bed just opened, — even large trees are growing. The beds are frequently covered with cranberry-bushes, hazel, willow, and the like, similar to the vegetation on old beaver meadows. In some countries, as in Ireland, parts of Switzerland, and Fuegia, in humid and quite cool climates, peat may form on plains, and even on hill-sides. This is called mountain or hill peat. In a few instances it has been found at the bottom of the sea, formed there under peculiar circumstances, from beds of sea-weed. The mountain peat, and that on plains, is of an inferior kind; and its beds usually do not exceed four feet in depth. We need not look for any in this region, except that formed in swamps or ponds, unless it be in low prairies. Our climate is usually too dry for that on plains or hill-sides.

The influence of heat, it may be mentioned, is so great, that peat is never found near the tropics. In the United States, it does not generally occur south of the Dismal Swamp. Vegetation, even under water, in hot climates, changes too fast and too completely to leave much carbon. In looking for beds in this region, we should examine only such places as are or have been quite depressed. Generally they will be found surrounded on all sides by higher land, making a basin-like depression, supplied with a moderate amount of water, and *generally underlaid with clay*, which, with the high land, prevents the water from either flowing or filtering away. Frequently the growth of moss and ferns in such places is favored by surrounding woods and the shade of hills. In such a place, if moss is growing plentifully, and the quantity of water is not too great, the probability is strong for a deposit of peat. The surface of beds, as stated, varies greatly, as is indicated by the various names in common use, such as turf, bog, moss, moor, and heath. Sometimes it is so wet, that the peat may be dipped out with a kind of scoop-net, as is occasionally done in Holland. The surface of others is firm enough to be crossed on foot, by stepping carefully on the more solid parts, the clumps of higher growth, or by jumping from hummock to hummock, — tussocks, as they are called in Scotland. Of others the surface is drier and more even, yet trembles as you pass over it: *quaking* bogs they are sometimes named in England and Scotland.

Animal remains in bogs are quite abundant, especially in Ireland, Flanders, and the Isle of Man; yet none have been found belonging to quadrupeds whose living species are peculiar to the tropics, such as the elephant, rhinoceros, hyena, tiger, hippopotamus, though remains of such are very common in other superficial deposits in Great Britain. These facts together demonstrate the geologically recent date of these bogs, as well as the existence of a tropical climate at an earlier time, even thus far north.

Usually peat-beds are very slow of growth. Occasionally, however, they increase with some rapidity. One case is mentioned of seven feet gain in thirty years. At the bottom of them, besides the clay mentioned, there are often deposits of *marl*, sometimes *shell marl*. This occurs when beds have grown in shallow lakes or ponds. Frequently these marl deposits are sufficient in amount to be used freely in agriculture. Occasionally thick crusts of *bog-iron* form on the bottom, and may be of much commercial value in some instances.

But how does peat look in the bed? The top layers, more generally called turf, are quite fibrous, loose, porous, full of leaves, roots, sticks, &c. Deeper down, it is more perfectly formed, pure, and dense. A few feet farther down, if good, it is quite free from roots and stems; and, if very good, is free from grit, stones, or other foreign matters. In color, it varies from yellowish and reddish and dark brown to pure black, called sometimes turbary bog.

In consistence, the deeper peat varies from thin mud to that of clay prepared for making brick; some so thin that it may be dipped up, while the drier peat will cut about like paste-blacking, and looking much like it. When thoroughly dry and pressed, it may appear like a rough cake of chocolate, with or without fibres in it, though under the knife it is much harder; or, if of better quality, it will look and cut like licorice-ball. Sometimes it crumbles when dry, but this is an indication that it is poor and imperfectly formed, too much like common muck. Usually when dry, if tolerably pure, it is quite firm and solid. When wet, if pure, the better qualities feel smooth, greasy soft, unctuous, like pure blue clay. Generally, though there are exceptions, it has but little odor, and but little taste. In weight it varies with the pressed and unpressed, the pure and impure specimens, from some so fibrous that it will even float on water, to others about the gravity of bituminous coal.

And now, knowing how it appears, and having reached the place, if you please, where we may reasonably hope to find it, how shall the examination be made?

If the surface is soft, run down a pole, and see if any comes up sticking to it. See if there is clay or marl on the end of the pole, if you have reached the bottom; or run down a gas-pipe, or a tin conductor from your eaves-trough, or a gun-barrel, and then push out what comes up inside; or tie a wide-mouthed vial to the end of your pole, with a string attached to the cork, and then, pulling out the cork at the depth you choose, bring it up, filled of course. If the bed is quite dry, dig off the top, and go down, first with the spade, then with the gas-pipe, &c. Be a Yankee, and you can examine it easily enough. When you have dried some, see how freely it will burn; how much ashes it leaves. It should burn in a stove or grate or fireplace, or heap out of doors, and without much smoke, if it is tolerably pure and dry.

It generally gives out, when burning, a bituminous odor, and sometimes, if quite turfy, the "peaty" odor, which some few will recognize, having observed it in their youth, when partaking of Irish or Scotch whiskey-punch, in the making of which whiskey, barley, malted with a peat-fire, is used, which thus communicates the peculiar flavor imitated here imperfectly with creosote.

Occasionally, in a very dry time, peat is discovered by fire which may have taken in it from a summer fallow, or from the woods. The peat is recognized from the burning continuing so long and deep. If you remember such a fire, it may be well to make examination. Very possibly, however, it will prove to be only muck, so poor in carbon that it will have but little value.

Next comes the question whether it is probable that much peat may be found, if properly prospected, in this country. Is peat of quite rare, or of frequent occurrence? A little study will show that it is found, and is still more frequently to be found, in vast amounts, and in very many places all through the Northern States.

Wood and coal, in this country, have been so cheap, that, thus far, peat has been discovered mostly by accident rather than by search, and is as little understood and valued as petroleum was but a few months — not years — ago. With us, its general usefulness and peculiar excellence for certain leading purposes have excited no proportionate interest. But all this is changing fast, and it will not be long before peat will stand prominent among the earth's great wealths; only a short time and, like iron and coal, it will be considered among the great natural accumulations of riches, and power, and national comfort, and well-being with which these Northern States are almost over-blessed.

Hundreds of beds are known, and thousands will be discovered, some in almost every county of the North. A word as to its abundance, first abroad, and then at home.

More than one tenth of Ireland is covered with peat; over 1,000,000 acres. One of the mosses on the Shannon is fifty miles in length by two or three in breadth. Beds vary from thin layers to ninety (90) feet in depth. In 1856, there were raised and registered in Ireland, England, and Scotland, 66,645,450 tons. The average depth of English beds is computed at 19½ feet. The bog of Montoire, near the mouth of the Loire, in France, is fifty leagues in circumference. On an average, there is employed in France the work of 50,000 men for forty days each year, in the cutting and preparation of peat. It is abundant throughout all Northern Europe.

As to the United States, though as yet so little looked for and so little used, it is already known to a great extent. A number of beds have lately been discovered within thirty miles of Watertown, and there are without doubt many more. The peat from some of these is apparently of the finest quality, pure, black, even, dense, free from fibres and impurities.

The Oswego bed, but lately opened, is about 270 acres in extent, varying from five to fifty feet in depth. Ten years ago even, it was estimated that in Massachusetts there were at least 120,000,000 cords. By the Massachusetts Geological Reports, it is mentioned in about one hundred towns; the beds being frequently underlaid with marl. It is abundant in New Jersey, in Connecticut, and Maine; in fact in all the New-England States. In New York, it is especially so along the Hudson River, — over 10,000 acres, it is stated, — and an immense supply in the central and western counties. One bed in Warren County is sixty feet in depth, and by report of Professor Emmons, must have been some 800 years in forming. Near Cold Spring a bed was discovered, the railroad track built over it having sunk in and disappeared.

In Washington County are over twenty beds; in Columbia County, thirty beds; in Dutchess County, forty beds; in Orange County are many beds: over 15,000 acres. The whole being over 30,000 acres. In twenty-eight of these beds marl is found. These last statistics, it should be remembered, are found in the New York Geological Survey, made twenty-two years ago, and the figures for the present would be much increased of course. Near Rome it is abundant, and near Little Falls, Cooperstown, Cherry Valley, Syracuse, &c., &c.: it will be found plentiful in St. Lawrence County and Lewis, and probably in Jefferson also. In Jefferson, there are some localities apparently well adapted to its growth; especially, perhaps, the towns of Theresa, Alexandria, Antwerp, Orleans, Pamelia, Philadelphia, Ellisburgh, Henderson, Cape Vincent, and Wilna. This, however, is by judging mainly from glancing over the map, and observing the distribution of low lands.

Now, what is it good for? and what is it worth? It is good in a stove, in a fireplace, in a furnace, for locomotives, — much used that way in England and Bavaria, — for stationary engines, wherever motive-power from fire is wanted, in the forge, in making pig-iron and steel, for puddling iron, for gas, as a fertilizer, for paraffine, &c., as mentioned below. It makes no clinkers in burning, nor are grate-bars consumed by it as by anthracite. The very best quality, well ground, pressed and dried, bulk for bulk, is nearly or quite equal to common bituminous coal for general purposes, and much superior for others. From this it goes down to almost worthless qualities. The coal, after the coking of some varieties, so it is stated, is preferable in making gunpowder to even that from dogwood and alder. Used as a fertilizer, it is sometimes first burned, at others unburned, mixed with manures, lime, &c., after it has been subjected to the frost and air. Its effects are said to be very permanent on land, lasting for years. In the manufacture of iron, especially in puddling iron, good qualities are far superior to mineral coal, and if first coked, is better even than charcoal. In Sweden, particularly, it is now employed for this purpose. Iron men will appreciate its excellence over anthracite in its being usually free from sulphur and phosphorus. It will probably be much used in this country before many years, both in blast and puddling furnaces, and perhaps in making steel.

Some varieties are of the very first value in manufacturing gas. The French have given this department especial attention.

Most good peat will produce much resin and paraffine, which last is quite valuable. It closely resembles spermaceti, and is used for candles, which burn like those from wax.

First-class dry peat, on analysis, averages about as follows to the 100 parts:—

Carbon	55
Hydrogen	5
Oxygen	35
Ashes	5
	100

The first three constitute the base from which the paraffine, oils, &c., are formed. (It should be mentioned, that in some instances, while peat is forming, part of the carbon of the woody fibre, among other chemical changes, retains hydrogen, and thus compounds result richer as hydrocarbons than the original fibre.) The very best peat contains about ninety per cent. of combustible matter.

Now, how is it prepared for use? This varies much with the service it is to perform. For ordinary fires, it is simply dried in the sun after being cut out in bricks with a *three-sided* spade. For some purposes it is pressed before drying. For other purposes it is ground previous to the pressing and drying. For others it is coked.

There are various machines for the grinding and pressing. The Boston Peat Company, whose office is 49 Congress Street, Boston, Leavitt & Hunnewell being the agents, use and also sell a machine said to be economical for beds even of an acre. It is reported, with ten-horse power (steam), to grind and press one hundred tons of crude peat, producing twenty-five tons or more of condensed fuel per day, nearly as solid as bituminous coal. This machine costs $1500; and as each bed, of course, furnishes itself with fuel gratis, the cost of working is stated to be about two dollars per ton.

As to the general market value of peat, at present but little can be said. Of prices at particular beds, the following are quotations: At the Oswego bed, pressed and dry, it goes at about $7 per ton, and delivered at Oswego, $7.50. At the Lexington bed mentioned, — or rather peat from it at Boston, — has heretofore usually sold for $10 per ton of 2000 pounds; but at present, if on hand, it would readily bring, by the thousands of tons, $12, — coal standing there at $15.

The cost of production and the price in market will hereafter, of course, vary exceedingly. A large bed, tolerably dry, or favorably situated for draining, located near a place of great consumption, and of pure, dense quality, may be worth anywhere from a thousand to some hundreds of thousands of dollars. If, however, badly located, very wet, quite poor, &c., it would be worth something to you if it was not on your farm.

There is one great advantage in looking for peat over boring for oil, for instance; for peat you put down your well with a pole, and it costs you only a couple of wet feet; and if you don't strike peat, peat has not struck you, — you are not floored; your financial eye is not closed up; you can go on looking; and if at last you see it, you can start business on a fair scale with only a spade or two; and thus this is a branch of industry eminently open to universally competitive enterprise, with small means and no indorser. How much capital, by the way, do you suppose is now, after only so few years' operations, engaged in petroleum? There are 1100 petroleum companies, with $100,000,000 of absolutely paid-in capital, $600,000,000 nominal. The product of this year, 1865, it is estimated will be 1,500,000 barrels, averaging about $14.50 per barrel.

Peat and petroleum (as their names indicate) are of the same family (Old King Coal being the head); and it would not be strange if the younger brother, with the shorter name, should do the longer and larger business after all, and become the greater favorite of the two.

There is more democracy in peat, more disposition to be useful generally, not distant at all, but at home on everybody's hearth; ready to lend a hand in the kitchen, or work in the shop, or start an engine, or run a factory, an iron-mill, if you call on him for a big job, anything 'most, and without much fuss and fixing; a good-natured, industrious, valid, and capable, rather "unwashed," rough sort of fellow, but a fellow substantial withal, plenty of means, and ready and able to do a good thing for you, and a big thing, too, if you want it. You had better make his acquaintance immediately if he happens in your neighborhood. R.

From "The Waltham Sentinel."

There is more in *peat* than is dreamed of in most men's philosophy.

FACTS ABOUT PEAT.

OPINIONS OF THE PRESS.

"An elaborate pamphlet, designed to show the economy of peat as a substitute for wood and coal, especially where fuel is required in large quantities. To this end, Mr. Leavitt has prepared an exhaustive statement of the history and properties of peat, the localities of peat-beds, the methods of preparation and manufacture, its applicability to the various arts, as well as to the production of heat, and other incidental matters of practical importance." — *Worcester Spy*.

"It will be found especially interesting to manufacturers and railroad managers." — *Hartford Courant*.

"It is full of information as to the value of peat, and its whereabouts." — *Bridgeport Farmer*.

"Mr. Leavitt shows conclusively, that peat is equally as good as coal, and for many purposes unquestionably superior, especially for generating steam, and for the manufacture of iron, steel, and other metals." — *Cheshire Republican*.

"The treatise is prepared with care, and embodies much useful information. The use of peat is now attracting, generally, the attention of railroad men, manufacturers, and others; and, in this view, the issue of the pamphlet is timely." — *Roxbury Journal*.

"An octavo pamphlet of 120 pages, elegantly printed, compiled by T. H. Leavitt, Esq. It is a thorough production, the author proceeding exhaustively, and arranging his abundant matter in a manner that renders the task of following him easy and profitable. He has mastered his subject, and evidently has neglected nothing that is calculated to illustrate it, and to press useful facts on the mind of the inquirer. Various, minute, and copious in its facts, and showing how valuable is peat as an article of fuel, this work must have a great effect in directing attention to a neglected agent for the production of heat, one which Providence has placed most freely at the command of man, and which ought to be made to enter very largely into human consumption. Mr. Leavitt is literally correct, when he says that the substance of which he treats so well 'is of sufficient importance to command earnest attention, not only from the business man, on the score of its application to domestic purposes, manufactures, and the arts, but from the philanthropist, in view of the relief it may be made to afford as one of the necessaries of life.' Mr. Leavitt's work should be read by all, as it is full of information, and it needs only that the value of peat should be understood, to bring it into general use, to the great relief of all interests." — *Boston Traveller*.

"The pamphlet is full of most interesting facts on the subject." — *Hartford Press.*

"Replete with interesting and instructive 'Facts,' demonstrating that abundant sources of supply are to be found in all the New England States, and its economy over wood and coal. We are persuaded its perusal by every consumer in New England would be productive of great good, and excite a new enterprise throughout the New England States." — *Portland Advertiser.*

"We would recommend any one whose pockets are interested to the amount of $25 for fuel, to get the book and read it." — *Brunswick, Me., Telegraph.*

"A well-timed and well-executed compilation of important facts. Of all men, farmers should turn their attention to peat for the preservation of their best forests for more valuable uses than fuel." — *Vermont Watchman.*

"The development of such a source of wealth and of general relief, lying all around, in lands that are otherwise utterly valueless, will add at once to the general resources of the country, and to the means of comfort and life of all classes of society. If this new method is what it is represented to be, of which we have satisfactory evidence, we earnestly hope it may attract the attention of capitalists and business men without delay." — *Brooklyn Union.*

"A very interesting work. The subject is beginning to excite considerable attention, not only on account of the fears of a growing scarcity of wood, its high price, and also that of coal, but because it forms one of the products of industry, which, when perfected from its raw state, forms, like mines of iron, lead, copper, and silver, *great* wealth to a nation." — *Union and Journal, Biddeford, Me.*

"A good-looking pamphlet of 120 pages, giving us all the information upon the subject of peat that the most laborious and extensive research can possibly furnish." — *Lowell Courier.*

"It contains more historical facts on the formation of peat-beds than any thing we have heretofore seen." — *New York Spirit of the Times.*

"A company in Boston have at length completed the invention of a simple and rational process, by which crude peat, as it is taken from its bed, can be converted into a solid, dry fuel, in good shape, in large quantities, and at a moderate cost. The machinery required is simple, and not too expensive for use, and can easily be set up and run by the side of the peat-bed. We have seen specimens of the peat condensed by this process, and are acquainted with the principle on which the machine works, and think there is no reason to doubt its efficiency. The personal character of the principal managers is fitted to inspire confidence that they would not come before the public, unless they had a good thing, calculated to be a general benefit. Indeed, their method may fairly claim to be not only the best, but the only one, so far as is known, in this country, that is at once effectual, cheap, and rapid. They are the pioneers of the present movement in favor of the use of peat, having been engaged for several years in their experiments and inquiries. Their actuary, Mr. T. H. Leavitt, has published a pamphlet of 118 pages, containing more information — historical, scientific, and practical — about peat, we venture to say, than any other man in the country is possessed of; and his perseverance is well entitled to the success which seems about to be realized.

"Those who seek to be thoroughly informed should procure it. And those who would transact business with the company should address the agents, Leavitt & Hunnewell, 49 Congress Street, Boston." — *Brooklyn Union.*

"A very handsomely printed and valuable pamphlet, containing 'Facts' which are not only interesting, but, if borne out by actual experiment, of the highest value. Turning listlessly to its title-page, we became so much absorbed in its pages, that we read it as closely as an editor ever finds time to read anything; and we came to the conclusion that the company who have undertaken to develop the value of peat ought to be encouraged by all who have money to invest in new, and probably remunerative channels of trade or manufactures. The glowing heat and cheerful light of a peat fire are the very ultimatum of a social evening." — *Providence Daily Press.*

"The work is deserving of more than passing notice. We commend it, and the subject of which it treats, to the attention of all; for no subject is of wider interest than that of fuel." — *Newport Daily News.*

"There is an abundance of it, thousands and millions of cords, scattered all over New England, and indeed throughout the North. It lies at our very doors. It is time that our people were stirring themselves to see if it may not be brought into general use, in order to stop the waste of our forests, now growing every year more valuable for timber, cheapen the cost of fuel, and render us less dependent upon coal monopolists and speculators for this indispensable article. The 'Facts,' compiled by Mr. Leavitt, embrace much interesting and valuable information." — *Boston Journal.*

"A glance over its pages shows that the compiler has labored with care and discrimination in the collection and presentation of his materials. We have a fondness for peat. The fires of love never burn dimly beside it. Its introduction as an article of fuel is no mere fancy, but a subject of grave importance." — *Brunswick Telegraph.*

"A thorough treatise on the qualities and practical uses of peat." — *Old Colony Memorial.*

"We consider it a valuable, timely, and interesting work. The whole community are interested in the subject of which it treats." — *Fall River News.*

"The questions of its supply, preparation, and most economical use, are of high interest; and this pamphlet embodies much needed information, which will aid in their solution. It embraces much curious and instructive matter of practical and scientific interest." — *Lowell Citizen.*

"Mr. Leavitt's facts and remarks throw a great amount of light upon the subject, and they ought to have a wide circulation. We have an abundance of peat; and the pamphlet tells us of its importance as an article of fuel, and how to prepare and use it." — *Pawtucket, R. I., Gazette.*

"A valuable pamphlet, containing 'Facts.'" — *Salem Mercury.*

www.ingramcontent.com/pod-product-compliance
Lightning Source LLC
LaVergne TN
LVHW020218110826
845151LV00003B/752

* 9 7 8 1 4 2 5 5 3 3 2 4 3 *